INTRODUCTION TO
DIGITAL LOGIC DESIGN

By Rajiv Kapadia

Minnesota State University - Mankato

cognella®
academic publishing

Bassim Hamadeh, CEO and Publisher
Michael Simpson, Vice President of Acquisitions
Jamie Giganti, Senior Managing Editor
Miguel Macias, Graphic Designer
John Remington, Senior Field Acquisitions Editor
Monika Dziamka, Project Editor
Brian Fahey, Licensing Specialist
Claire Yee, Interior Designer

First published in the United States of America in 2016 by Cognella, Inc.

Trademark Notice: Product or corporate names may be trademarks or registered trademarks, and are used only for identification and explanation without intent to infringe.

Cover image copyright© by Depositphotos / Raimundas.

Printed in the United States of America

ISBN: 978-1-63487-320-8 (pbk) / 978-1-63487-321-5 (br)

❀ **cognella**®
academic publishing

www.cognella.com   800-200-3908

TABLE OF CONTENTS

DEDICATION

My Mother, without whose encouragement and
support nothing would have been possible.

My Children, Ajay and Tara, whose help in
completing the manuscript cannot be measured.

I. PREFACE

This book was written after teaching Introduction to Digital Logic Design to freshmen students who come to college right out of high school. These students need a lot of "hand holding," and this book provides the proper mix of hand holding and in-depth work needed for understanding the subject area of digital logic design.

PEDAGOGY

The contents of the book are laid out in the manner in which the material is typically covered over the course of one semester. This layout supports the author's lectures and helps the students to build the foundation necessary for further studies. All the important concepts and conclusions have supporting drill exercises (with answers for the students), and there are additional homework problems at the end of each chapter.

ACCURACY

The goal of any educational publication is to be free of errors. The absolute worst thing that a student will find in a textbook is errors, so all the people involved in completing and printing this book are absolutely dedicated to making sure that there are no errors in the book.

USE OF THE TEXT

In general, the text is written so that the instructor will begin with the first chapter and go sequentially to the last chapter. The instructor should be able to complete the first two chapters in the first two weeks of class, with each week having three class meetings. Chapters 3 and 4 will require more time, as there are procedures that have to be mastered by the student. For this reason, each of these chapters will require two weeks to complete. Chapters 5 and 6 are shorter chapters; together they should require approximately three weeks to complete. The last two chapters are a little challenging for the students. The time spent on these chapters will pay a handsome reward if the instructor does not have to hurry to complete these two chapters in less than four weeks. With this timeline, there is enough time available for midterm exams and a review at the end of the semester.

I. NUMBER SYSTEMS AND CODES

1.0.   INTRODUCTION TO THE BINARY NUMBER SYSTEM

In this class, we will investigate what makes digital components function as they do. As we go through the semester, we will see how digital components are used to control functions, represent data, draw pictures, represent numbers, and numerous other uses. To do all this, the digital systems operate on a binary system. A binary system has only two distinct states. The states a binary system uses can be represented in many different ways; generally, we use the "ON" and "OFF" states or the "0" and "1" states. In this book, we will use the "0" and "1" states most of the time. In this chapter, we will begin by representing the numbers in a binary system; next, we will see how we can convert between binary and decimal number systems. We will end the chapter by looking at other binary codes.

1.1.   THE POSITION NUMBER SYSTEM

In a position number system, the value of a symbol is represented by its numerical value multiplied by its positional value. The positional value of any digit is interpreted as shown in Equation 1.1.

$$Positional\ value = Base^{(position)}$$

(1.1)

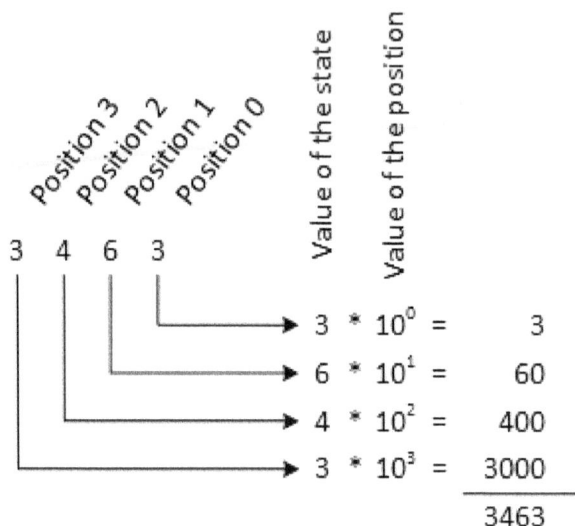

FIGURE 1.1. Interpreting the value of a base 10 number.

In Equation 1.1, the base represents the number of different symbols that a number system uses to represent any number. The position represents where the number is located relative to the point that separates the fraction part of the number and the integer part of the number. We call this the "radix point"; in a decimal number system, we refer to this as the decimal point. To count the position, we begin from the radix point, and the first digit to the left of the radix point is in position zero. From there, we count positive position values to the left and negative position values to the right. So, for example, in the decimal number 1235.73, the digit 2 is in position 2, while the digit 7 is in position −1.u symbols that the system uses to represent any number. In Equation 1.1 above, we use the base to determine the positional value of any digit. With only these ten states available to represent larger numbers, we assign to each state the magnitude value and its positional value. We can determine the value of any number using the magnitude and positional value, as shown in Figure 1.1 and in Equation 1.2.

$$Value\ of\ any\ digit = Value\ of\ the\ state * Value\ of\ the\ position \quad (1.2)$$

In Figure 1.1, we see that the digit 3 is used twice but its value is different each time it is used. In position zero, its value is just three, while in position 3,

INTRODUCTION TO DIGITAL LOGIC DESIGN

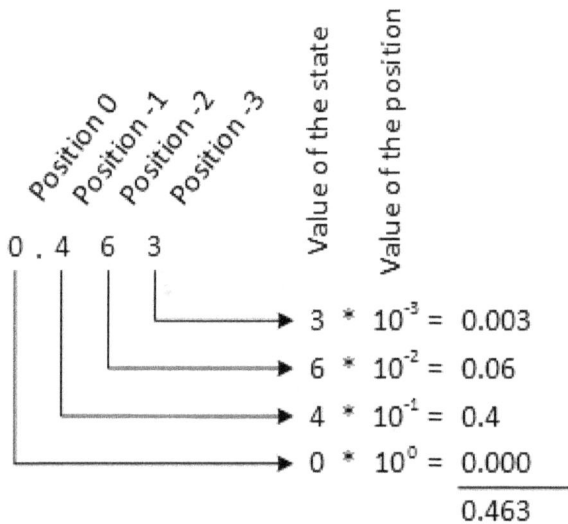

FIGURE 1.2. Interpreting the value of a fraction in base 10.

its value is three thousand. The value of any digit in a number can be determined in the same way. We use Equation 1.2 as shown in Figure 1.2.

1.1.1. REPRESENTING FRACTIONS IN THE DECIMAL NUMBER SYSTEM

To determine the value of digits that are fractional (or less than 1), we continue in the same manner except the position is now negative. Look at Figure 1.2; it shows how we assign value to fractions in a positional number system in general and the decimal number system in particular. The position of any digit is measured from the point that separates the integer portion of a number from the fraction portion of a number. The fractions are all in negative positions, and the integers are all in positive positions. This is the method used for all positional number systems. In Figure 1.2, the digit 6 is in position −2, so its value is $6 * 10^{-2} = 0.06$. In the same way, the positional value of all the digits can be determined.

1.1.2. THE BINARY NUMBER SYSTEM

In the previous section, we saw that the decimal number system is a positional number system with a base of 10. The binary number system is also a positional number system with a base 2. The binary number system has a base of 2 because it uses only two different symbols to represent all the numbers. These two symbols are 0 and 1. We can interpret the binary numbers in the same way we interpreted the decimal numbers in Figure 1.1 and Figure 1.2. This is shown in Figure 1.3. The difference between the two number systems is the base of the number system. Looking at Figure 1.3, we see how we can interpret a binary number in terms of its decimal equivalent. Notice that since the base is 2, the value of the position is based upon this number. The positions of the digits are counted in the same way as we did for the decimal number system. In Figure 1.3, the integer part of the number is 1011. In this integer, the zero is in position 2, and the one to its left is in position 3. The value of this 1 is $1 * 2^3 = 8$, as shown in Figure 1.3. Similarly, the zero in the fraction part is in position -1,

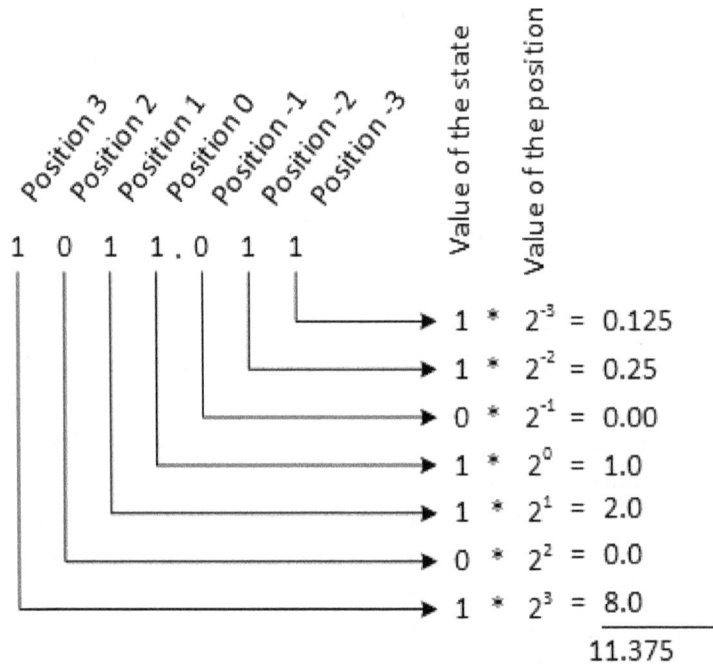

FIGURE 1.3. Interpreting the value of a number in base 2.

and the one to the right of it is in position −2. The value of this 1 is $1 * 2^{-2} = 0.25$, as shown in Figure 1.3.

1.1.3. CONVERTING FROM A NUMBER SYSTEM OF BASE N TO A DECIMAL NUMBER

In the previous section, we saw how we can convert a binary number to a base 10 number. We can use the same process to convert a number from a number system of any base to a number in the base 10 number system. To complete the conversion, all we need to adjust is the position value of the number. The following example will show how this is done. Convert $(632.45)_7$, a base 7 number, to a decimal number. This is shown in Figure 1.4. Notice first that a number in base 7 will be represented by digits 0, 1, 2, 3, 4, 5, and 6 and the position value of the digits are powers of 7.

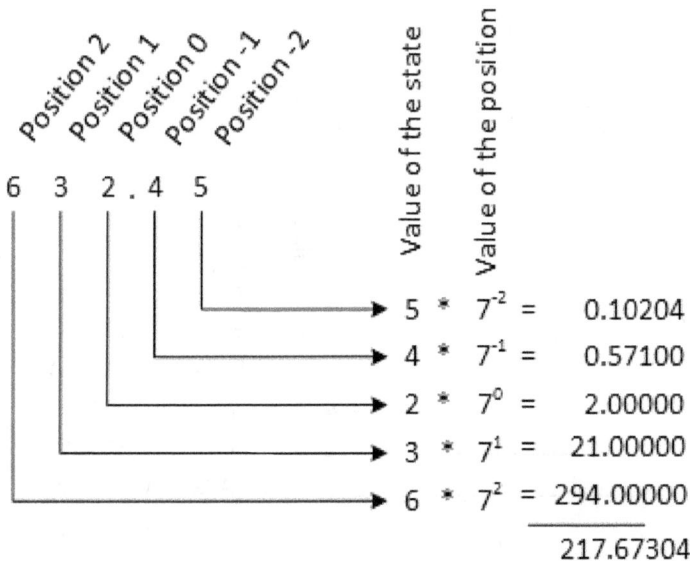

$5 * 7^{-2} =$	0.10204
$4 * 7^{-1} =$	0.57100
$2 * 7^{0} =$	2.00000
$3 * 7^{1} =$	21.00000
$6 * 7^{2} =$	294.00000
	217.67304

FIGURE 1.4. Interpreting the value of a number in base 7.

Review Questions for Section 1.1

Question: In the decimal number 6345.0134, what is the position value of the digit 6? What are the positional values of the two 3's? What are the positional values of the two 4's?

    Answer: The position value of 6 in position 3 → 1000. The position value of 3 in position 2 → 100. The position value of 3 in position −3 → 0.001. The position value of 4 in position 2 → 10. The position value of 4 in position −4 → 0.0001.

Question: In the binary number 1010.011, what are the position values of the 1 digits? What are the positional values of all the 0 digits?

    Answer: The position value of 1 in position 3 → 8. The position value of 1 in position 1 → 1. The position value of 1 in position −2 → 0.25. The position value of 1 in position −3 → 0.125. The position value of 0 in position 2 → 0. The position value of 0 in position 0 → 0. The position value of 0 in position −1 → 0.

Question: What is the decimal equivalent of the number $(726.43)_9$ if this number is a base 9 number?

    Answer: Examine Figure 1.5. It shows how we convert from base 9 to a decimal number system.

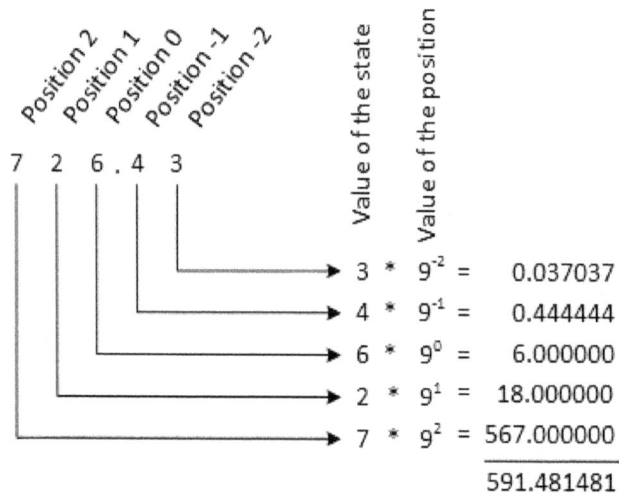

FIGURE 1.5. Interpreting the value of a number in base 9.

1.2. CONVERTING FROM ONE NUMBER SYSTEM TO ANOTHER

The previous section showed us how to determine the decimal equivalent of a number in any base. Next, we will examine how to convert a number from base 10 system to a number in a different base. The process is exactly opposite to the process that we have just seen in the previous section. We break this process into converting integers first and then the fractional part, as the two procedures are slightly different.

1.2.1. CONVERTING INTEGERS FROM DECIMAL TO A DIFFERENT N NUMBER SYSTEM

In Figure 1.6, we convert the number 748, which is a base 10 number, to a number in base 5. We begin by dividing the base 10 number with the base of the number we want. In this example, we want to convert the base 10 number to a base 5 number, so we divide the base 10 number by 5. The division will give us an integer result and a remainder. We place the remainder to the side and continue to work with the integer result. In Figure 1.6, we have continued the division of the integer result portion and placed the remainder to the side. Notice that the remainder is written as a whole number and not a fraction that you would get if you were using your calculator to complete the division process. We continue the division of the integer result and write the remainder to the side until the integer result from the division process is zero. We are now ready to write the number in the new base. The most

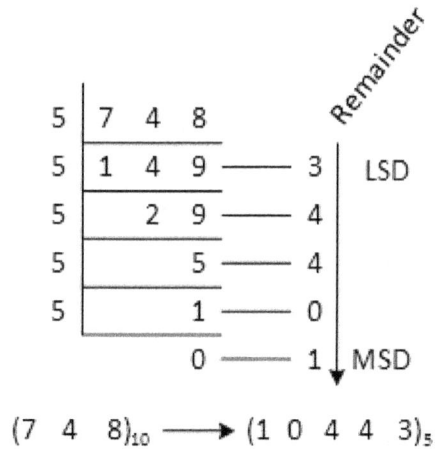

$$(7 \quad 4 \quad 8)_{10} \longrightarrow (1 \quad 0 \quad 4 \quad 4 \quad 3)_5$$

FIGURE 1.6. Converting an integer from Base 10 to a number in a different base.

significant digit of the number in the new base is the result of the remainder of the last division. The least significant digit is the remainder left over from the first division process. So reading the number from the most significant digit to the least significant digit will give us the new number in the required base that is equivalent to the number in base 10. So for our example, we will have $(748)_{10} = (10443)_5$. You can verify the result by converting $(10443)_5$ to a number in base 10; you should get $(748)_{10}$.

1.2.2. CONVERTING FRACTIONS FROM DECIMAL TO A BASE N NUMBER SYSTEM

To convert the fraction portion of the number, the process is slightly different. Instead of dividing the fraction by the base of the number system, we multiply the fraction by the base of the number system. Then instead of keeping the fraction portion, we put aside the integer portion of the result and continue to work with the fraction portion of the result. Examine Figure 1.7, where we convert a fractional number from base 10 to a number in base 5.

In Figure 1.7, we want to convert the fraction $(0.904)_{10}$ from base 10 to a fraction in base 5. To do this, we first multiply the fraction in base 10 with the base of the target number system. This will give us an integer portion and a fraction portion. In our example, the integer portion is 4. We set the integer aside and continue to work with the remaining fraction, which in our example is 0.52. The process is to set the integer portion aside and work with the fraction. Again we multiply this new fraction by the base of the target number system to get an integer and a

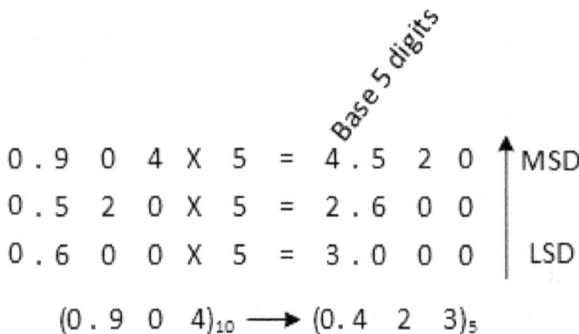

```
                                    Base 5 digits
0 . 9  0  4  X  5  =  4 . 5  2  0   ↑ MSD
0 . 5  2  0  X  5  =  2 . 6  0  0   |
0 . 6  0  0  X  5  =  3 . 0  0  0   |  LSD

     (0 . 9  0  4)₁₀  ──►  (0. 4  2  3)₅
```

FIGURE 1.7. Converting a fraction from Base 10 to a number in a different base.

fraction. We put the integer aside and work with the fraction. We continue this process until we have enough significant digits. In the diagram, we see that $(0.904)_{10} = (0.423)_5$. This process will not always end as shown in Figure 1.7. It may continue forever. So as a general rule, if we have m digits after the decimal point, we continue the conversion until it ends or until we have one more digit after the decimal point. Figure 1.8 shows us how we use this terminating rule.

We see in Figure 1.8 that the process has not terminated even after we have obtained five digits in the new base. This is often the case when we are converting fractions, so we arbitrarily decide to terminate the process after we have one more digit after the decimal point in the new base than we had in the old base.

1.2.3. CONVERTING NUMBERS FROM BASE M TO A BASE N NUMBER SYSTEM

We have now seen how we can convert a number from a base N number system to a base 10 number system in Section 1.1. We have also seen how we can convert a number from a base 10 number system to a number system

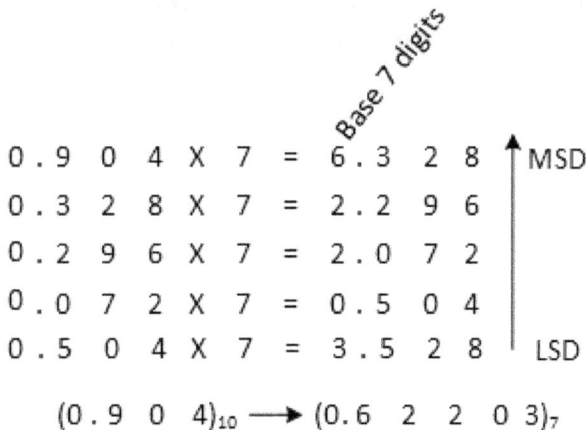

Base 7 digits

$$
\begin{array}{l}
0.9\ 0\ 4 \times 7 = 6.3\ 2\ 8 \quad \uparrow \text{MSD} \\
0.3\ 2\ 8 \times 7 = 2.2\ 9\ 6 \\
0.2\ 9\ 6 \times 7 = 2.0\ 7\ 2 \\
0.0\ 7\ 2 \times 7 = 0.5\ 0\ 4 \\
0.5\ 0\ 4 \times 7 = 3.5\ 2\ 8 \quad \text{LSD}
\end{array}
$$

$$(0.9\ 0\ 4)_{10} \longrightarrow (0.6\ 2\ 2\ 0\ 3)_7$$

FIGURE 1.8. Converting a fraction from Base 10 to a number in a different base.

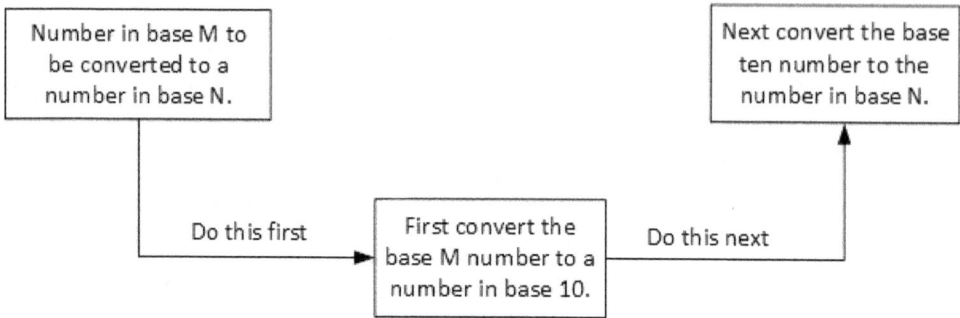

FIGURE 1.9. Converting a number from base M to a number in base N.

in base N. We have done this conversion with the base 10 number system as one of the number systems because we know how to do arithmetic in the base 10 number system. So when we want to convert a number from base M to a number in base N number system, we take a detour to the base 10 system, as shown in Figure 1.9.

We have already seen how the two individual conversions are completed. It just remains for us to put the two of them together one after the other. One example will make that clear. Say we want to convert $(372.45)_8$ to a number in base 9. We first begin by converting the base 8 number to a base 10 number. This is shown in Figure 1.10. This completes the conversion from base 8 to base 10. Since our target base is base 9, we convert this base 10 number to a number in base 9. This is shown in Figure 1.10. This two-step process will convert a number from any base to a number in any other base.

```
3  7  2 . 4  5
```

$$5 * 8^{-2} = \quad 0.078125$$
$$4 * 8^{-1} = \quad 0.500000$$
$$2 * 8^{0} = \quad 2.000000$$
$$7 * 8^{1} = \quad 56.000000$$
$$3 * 8^{2} = 192.000000$$
$$\overline{250.578125}$$

$$(3\ 7\ 2\ .\ 4\ 5)_8 \longrightarrow (2\ 5\ 0\ .5\ 7\ 8)_{10}$$

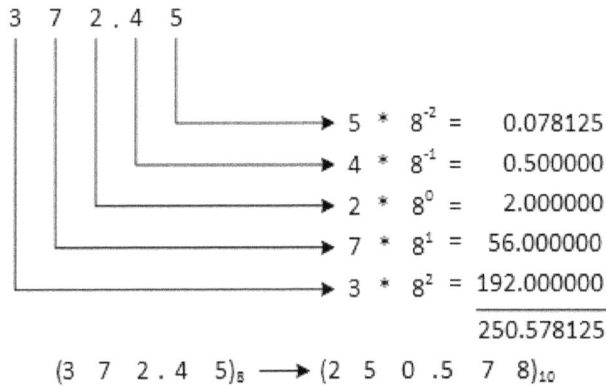

Figure 1.10 a First convert a base 8 number to a base ten number.

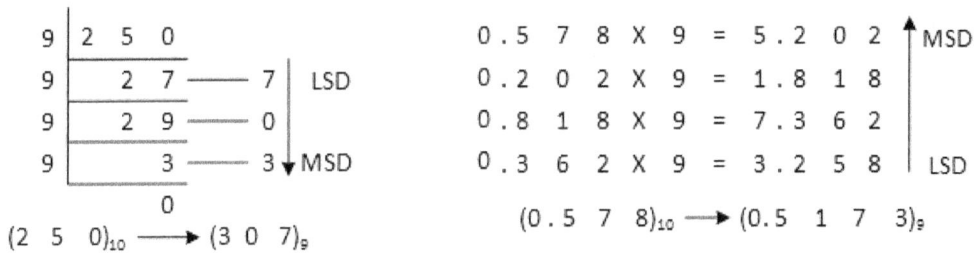

```
9 | 2 5 0                    0 . 5  7  8  X  9  =  5 . 2  0  2   ↑ MSD
9 |   2 7 ──── 7 | LSD       0 . 2  0  2  X  9  =  1 . 8  1  8   |
9 |   2 9 ──── 0 |           0 . 8  1  8  X  9  =  7 . 3  6  2   |
9 |     3 ──── 3 ↓ MSD       0 . 3  6  2  X  9  =  3 . 2  5  8   | LSD
        0
```

$$(2\ 5\ 0)_{10} \longrightarrow (3\ 0\ 7)_9 \qquad\qquad (0 . 5\ 7\ 8)_{10} \longrightarrow (0.5\ 1\ 7\ 3)_9$$

Figure 1.10 b Next converting the base 10 number to a base nine
number. Note the integer and the fractions are converted individually.

$$(3\ 7\ 2\ .\ 4\ 5)_8 \longrightarrow (2\ 5\ 0\ .5\ 7\ 8)_{10} \longrightarrow (3\ 0\ 7\ .5\ 1\ 7\ 3)_9$$

FIGURE 1.10. Two step conversion to convert a base M number to a base N number.

Review Questions for Section 1.2

Question: Convert the decimal number 6235.0134 to a number in base 3.
  Answer: The integer portion of the number is converted by repeatedly
  dividing (6235) by 3 to get the integer portion in base 3 as $(22112212)_3$,
  and the fraction portion is obtained by repeatedly multiplying the fraction portion by the new base 3 to get $(0.0001002)_3$. So the complete
  number when converted will be as shown in Equation 1.3.

$$(6235.0134)_{10} \rightarrow (22112212.0001002)_3 \qquad\qquad (1.3)$$

Question: Convert the number $(472.638)_8$ to a number in base 5.

Answer: The number is first converted to a base 10 number by multiplying each digit with its positional value to get a number in base 10. This gives us a base 10 number, which is $(314.8125)_{10}$. Next, we convert this base 10 number to a number in base 5 by first dividing the integer portion by 5 and saving the remainders, as shown in Figure 1.6. When the integer portion is completed, we should get $(2224)_5$. To convert the fraction portion, we multiply the base 10 fraction by 5 and save the integer portion. For our example, we will get $(0.4012)_5$. Thus, the complete conversion can be written as

$$(472.638)_8 \leftrightarrow (314.8125)_{10} \leftrightarrow (2224.4012)_5$$

1.3. THE OCTAL AND THE HEXA-DECIMAL NUMBER SYSTEM

The conversion process for converting a number from one number system to another number system as seen in Section 1.2 works just fine no matter what the two bases are. There is a much simpler method if a special relation like the one shown in Equation 1.4 relates the two bases to each other. When an exponential relation relates the two bases to each other where one base is a power of the other base, we can use a rather simple procedure. For example, assume that the two bases α and β are related by the exponential relation, as shown in Equation 1.4.

$$\alpha^3 = \beta \tag{1.4}$$

With such a relation, three digits in the number in base α can represent one digit in the number with base β. So if we want to convert a number from base α to a number in base β, we group all the digits in base α in groups of three starting from the radix point. Then we convert each group of digits to get the number in base β. Here, we examine two special relations, as these two relations often come up when we are working with digital logic and digital circuits such as the computer. Say the base α is 2 and the base β is 8. Then the conversion can be accomplished by looking at the relation given in Figure 1.11. According to Figure 1.11, any time we see the group $(101)_2$, we can replace it with the digit $(5)_8$, and so on.

1.3.1. RELATION BETWEEN BINARY AND OCTAL NUMBER SYSTEM

Base 2 numbers			Base 8 numbers
0	0	0	0
0	0	1	1
0	1	0	2
0	1	1	3
1	0	0	4
1	0	1	5
1	1	0	6
1	1	1	7

FIGURE 1.11. Relation between numbers in two different bases when one is an exponential power of the other.

The first such relation that we will examine is the relation between the binary and the octal number system. Examine Figure 1.11; there, we see that every base 8 digit can be represented by three base 2 digits. It is due to this special relation that we can convert between the two bases by looking up a table like the one shown in Figure 1.11. With the table in Figure 1.11, we can now convert either from base 2 to base 8 or from base 8 to base 2. The following examples will make the conversion process clear. As a first example, we want to convert $(1101010.010110)_2$ to a number in base 8, as shown in Equation 1.5.

$$\underset{\text{Group}}{1}\,\underset{\text{Group}}{101}\,\underset{\text{Group}}{010}.\underset{\text{Group}}{010}\,\underset{\text{Group}}{110} = (152.26)_8 \tag{1.5}$$

To complete the required conversion, we started to form groups of three digits in the base 2 number. Notice we always start making these groups from the binary (radix) point, going left to group the integers and going right to group the fraction portion. Next, we replaced each group in the base 2 number by the corresponding digit from the table that shows the relation between the base 2 digits and the base 8 digits. We can use a similar process to convert a base 8 number to a base 2 number, as shown in Equation 1.6.

$$(427.64)_8 = \underbrace{100}_{Group}\ \underbrace{010}_{Group}\ \underbrace{111}_{Group}.\underbrace{110}_{Group}\ \underbrace{100}_{Group} \tag{1.6}$$

This section shows that converting numbers between base 2 and base 8 is a simple matter of table lookup.

1.3.2. RELATION BETWEEN BINARY AND HEXA-DECIMAL NUMBER SYSTEM

Binary Digit	Hex Digit
0000	0
0001	1
0010	2
0011	3
0100	4
0101	5
0110	6
0111	7
1000	8
1001	9
1010	A
1011	B
1100	C
1101	D
1110	E
1111	F

FIGURE 1.12. Relation between the Binary number system and the Hexa-decimal number system.

The relation between the binary and the hexa-decimal number systems is also an exponential relation. Each hexa-decimal digit can be represented by four binary digits since the relation between the two number systems can be written as shown in Equation 1.7; the table to convert binary digits to hexa-decimal digits is shown in Figure 1.12.

$$2^4 = 16 \tag{1.7}$$

Examine Figure 1.12; there, we see the hexa-decimal number system. Since the hexa-decimal number system needs sixteen unique symbols to represent all the possible digits, it is customary to use the ten digits of the decimal number system and then use the first six letters of the alphabet to make up the remaining digits. The hexa-decimal number system has to represent sixteen unique digits. Each one of the sixteen digits is represented by four base 2 digits. The relation is shown in Figure 1.12. The following examples will make the conversion process clear. As a first example, we want to convert $(110110110010.01001011)_2$ to a number in base 16, as shown in Equation 1.8. To complete the required conversion, we started to form groups of four digits in the base 2 number. We always start making these groups from the binary point, going left to group the integers and going right to group the fraction portion.

$$\underbrace{1101}_{Group}\,\underbrace{1011}_{Group}\,\underbrace{0010}_{Group}\,.\,\underbrace{0100}_{Group}\,\underbrace{1011}_{Group} = \left(DB2.4B\right)_{16} \qquad (1.8)$$

After we have formed the groups, we replaced each group in the base 2 number by the corresponding digit from the table that shows the relation between the base 2 digits and the base 16 digits. We can use a similar process to convert a base 16 number to a base 2 number, as shown in Equation 1.9.

$$\left(269.24\right)_{16} = \underbrace{10}_{Group}\,\underbrace{0110}_{Group}\,\underbrace{1001}_{Group}\,.\,\underbrace{0010}_{Group}\,\underbrace{0100}_{Group} \qquad (1.9)$$

This section shows that converting numbers between base 2 and base 16 is a simple matter of table lookup.

Review Questions for Section 1.3

Question: Convert the binary number 10010110101.011011 to a number in base 8 and a number in base 16 using the relation explained in Section 1.3.

    Answer: First, we convert the binary number to the base 8 number, as shown in Equation 1.10. Next, we convert the number to base 16. This is shown in Equation 1.11.

$$\underbrace{10}_{Group}\,\underbrace{010}_{Group}\,\underbrace{110}_{Group}\,\underbrace{101}_{Group}\,.\,\underbrace{011}_{Group}\,\underbrace{011}_{Group} = \left(2265.33\right)_{8} \qquad (1.10)$$

$$\underbrace{100}_{Group}\,\underbrace{1011}_{Group}\,\underbrace{0101}_{Group}\,.\,\underbrace{0110}_{Group}\,\underbrace{11}_{Group} = \left(4B5.6C\right)_{16} \qquad (1.11)$$

Question: Convert the base 8 number $(1426.357)_{8}$ to a number in base 2 using the relation explained in Section 1.3.

    Answer: This time, we convert each base 8 digit into three digits of the binary number, as shown in Equation 1.12.

$$\left(1426.357\right)_{8} = \underbrace{1}_{Group}\,\underbrace{100}_{Group}\,\underbrace{010}_{Group}\,\underbrace{110}_{Group}\,.\,\underbrace{011}_{Group}\,\underbrace{101}_{Group}\,\underbrace{111}_{Group} \qquad (1.12)$$

Question: Convert the base 16 number $(149.A57)_{16}$ to a number in base 2 using the relation explained in Section 1.3.

    Answer: This time, we convert each base 16 digit into four digits of the binary number, as shown in Equation 1.13.

$$(149.A57)_8 = \underbrace{1}_{Group}\ \underbrace{0100}_{Group}\underbrace{1001}_{Group}.\underbrace{1010}_{Group}\underbrace{0101}_{Group}\underbrace{0111}_{Group} \qquad (1.13)$$

1.4. CHAPTER PROBLEMS

1.4.1.   For a base 10 number system, what is the positional value of a digit that is three places to the right of the decimal point? To the left of the decimal point?

1.4.2.   For a base 10 number system, what is the positional value of a digit that is 0.01? Where is this digit located in the number?

1.4.3.   For a base 6 number system, what is the positional value of a digit that is three places to the right of the decimal point? To the left of the decimal point?

1.4.4.   For a base 6 number system, what is the positional value of a digit that is 216? Where is this digit located in the number?

1.4.5.   Convert the number $(385.236)10$ to a number in base 2 number.

1.4.6.   Convert the number $(385.236)10$ to a number in base 3 number.

1.4.7.   Convert the number $(385.236)10$ to a number in base 5 number.

1.4.8.   Convert the number $(385.236)10$ to a number in base 7 number.

1.4.9.   Convert the number $(345.236)8$ to a number in base 10 number.

1.4.10. Convert the number $(5.23)6$ to a number in base 10 number.

1.4.11. Convert the number $(102.2)3$ to a number in base 10 number.

1.4.12. Convert the number $(2256.214)9$ to a number in base 10 number.

1.4.13. Convert the number $(345.236)8$ to a number in base 6 number.

1.4.14. Convert the number $(5.23)6$ to a number in base 4 number.

1.4.15. Convert the number $(102.2)3$ to a number in base 8 number.

1.4.16. Convert the number $(2256.214)9$ to a number in base 7 number.

1.4.17. Convert the number (1011.1101)2 to a number in base 4 number using the grouping technique.

1.4.18. Convert the number (1010101.0110101)2 to a number in base 8 number using the grouping technique.

1.4.19. Convert the number (101011.11011)2 to a number in base 4 number using the grouping technique.

1.4.20. Convert the number (101101101.110011)2 to a number in base 8 number using the grouping technique.

2.   SIMPLE LOGIC GATES

2.0. THE FUNDAMENTAL LOGIC GATES

The logic gate represents the simplest building block of any digital system. If you are given enough logic gates of different types, you can build an iPhone, a computer, or the tablet that you have at home; in fact, you can build any digital system that you know of. This means that all digital systems use logic gates, and it is for this reason that we start our study of digital systems with the study of logic gates. The logic gate can be thought of in one of two different ways. In its simplest form, we can think of the logic gate as an input-output system. The logic gate accepts one or more logic signals as input; it then combines these logic signals according to some logic rule and provides one output. Each logic gate uses a different rule. Another way to think of the logic gate is as a switch and a rule. In this way of thinking, we say that the switch will open if the inputs to the gate obey the rule. If the inputs to the gate violate the rule, then the switch will be closed. Thinking of the logic gate as a switch allows us to assign logical values to the output of the logic gate, which can be treated as logic High (switch closed) or logic Low (switch open). Sometimes, we say that the output of the logic gate is either 1 or 0. This is the same as thinking of the logic gate as a switch. In this interpretation, we say that when the switch is closed, the output is high or 1, and when the switch is open, the output is low or 0. Since we are using 0's and 1's to represent when the logic gate obeys or violates the rule, we will define the rule in terms of a table that we will call the "truth table." We will use the concept of the truth table to define not only how a logic gate behaves, but to define how any logic circuit behaves. There are five basic logic gates and a couple of derived logic gates. In this chapter, we look at the logic gates and see how we can use them to build a logic function.

2.1. THE AND GATE

The first logic gate that we will examine is a two-input AND gate. To learn how to use the gate, we need to look at the function that the gate represents. The AND gate can be represented in terms of a logic equation and the truth table for the logic gate. Look at Figure 2.1; there, we show the diagram that is used to represent the AND gate in circuits and the truth table that the AND gate obeys. The function that the AND gate obeys is represented as an equation in Equation 2.1. Looking at the logic equation, we read Equation 2.1 as follows: When input X is logic High AND input Y is logic High, then the output from the AND gate is logic High. When either of the two inputs is logic Low, then the output from the AND gate is logic Low. Notice that the " $\bullet$ " symbol is used to represent the logic function "AND," so when you are reading a logic equation, you must convert the " $\bullet$ " symbol to the "AND" operation and not the product operation.

$$F = X \bullet Y \qquad (2.1)$$

This same information is present in the truth table for the logic gate. Let us see how this works. The first line in the truth table reads as: When input X is Low and input Y is Low, then the output F is Low. We read the second line as: When input X is Low and input Y is High, then the output F is Low. We read the third as: When input X is High and input Y is Low, then the output F is Low. We read the last line as: When input X is High and input Y is High,

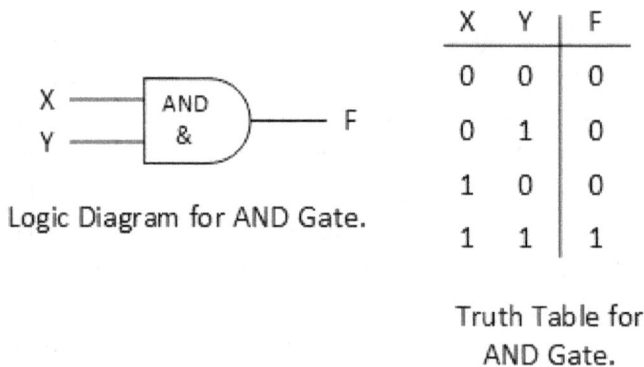

X	Y	F
0	0	0
0	1	0
1	0	0
1	1	1

Logic Diagram for AND Gate.

Truth Table for AND Gate.

FIGURE 2.1. Logic Diagram and Truth Table for the AND Gate.

then the output F is High. Thus, both the truth table and the logic equation convey the same information.

2.2.   THE OR GATE

The next logic gate that we will examine is a two-input OR gate. To learn how to use the gate, we need to look at the function that the gate represents. The OR gate can be represented in terms of a logic equation and the truth table for the logic gate. Look at Figure 2.2; there, we show the diagram that is used to represent the OR gate in circuits and the truth table that the OR gate obeys. The function that the OR gate obeys is represented as an equation in Equation 2.2. Looking at the logic equation, we read Equation 2.2 as follows: When input X is logic High OR input Y is logic High, then the output from the OR gate is logic High. When both the inputs are logic Low, then the output from the OR gate is logic Low. Notice that the "+" symbol is used to represent the logic function "Or," so when you are reading a logic equation, you must convert the "+" symbol to the "Or" operation and not the addition operation.

$$F = X + Y \qquad (2.2)$$

This same information is present in the truth table for the logic gate. Let us see how this works. The first line in the truth table reads as: When input X

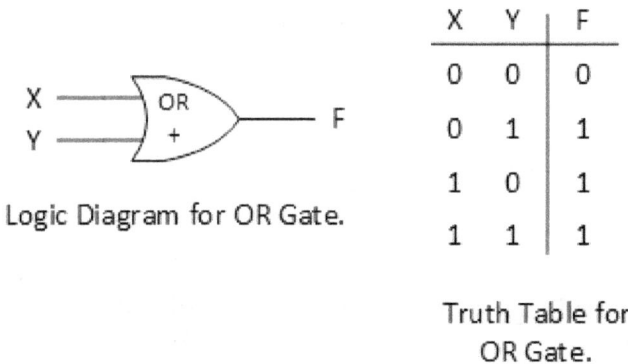

X	Y	F
0	0	0
0	1	1
1	0	1
1	1	1

Logic Diagram for OR Gate.

Truth Table for OR Gate.

FIGURE 2.2. Logic Diagram and Truth Table for the OR Gate.

is Low and input Y is Low, then the output F is Low. We read the second line as: When input X is Low and input Y is High, then the output F is High. We read the third as: When input X is High and input Y is Low, then the output F is High. We read the last line as: When input X is High and input Y is High, then the output F is High. Thus, both the truth table and the logic equation convey the same information. When any one of the two inputs is High, then the output is going to be High.

2.3.   THE NOT GATE

The next logic gate that we will examine is a NOT gate; this is a single input gate. To learn how to use the gate, we need to look at the function that the gate represents. The NOT gate can be represented in terms of a logic equation and the truth table for the logic gate. Look at Figure 2.3; there, we show the diagram that is used to represent the NOT gate in circuits and the truth table that the NOT gate obeys. The function that the NOT gate obeys is represented as an equation in Equation 2.3. Looking at the logic equation, we read Equation 2.3 as follows: When input X is logic High, then the output from the NOT gate is logic Low. The NOT gate complements the input to give us the output. To represent the complement, we either draw a bar over the variable that is complemented or write the exclamation mark before the variable that is complemented. We read Equation 2.3 as "F equals X bar" or as "F equals Not X."

$$F = \bar{X} = !X \qquad\qquad (2.3)$$

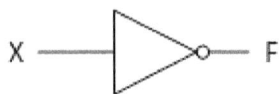

X	F
0	1
1	0

Logic Diagram for NOT Gate.   Truth Table for NOT Gate.

FIGURE 2.3. Logic Diagram and Truth Table for the NOT Gate.

INTRODUCTION TO DIGITAL LOGIC DESIGN

This same information is present in the truth table for the logic gate. Let us see how this works. The first line in the truth table reads as: When input X is Low, output F is High. We read the second line as: When input X is High and input Y is High, then the output F is Low. Thus, we can say that the output of the NOT gate is the complement of the input.

2.4.   THE NAND GATE

The next logic gate that we will examine is a two-input NAND gate. To learn how to use the gate, we need to look at the function that the gate represents. The NAND gate can be represented in terms of a logic equation and the truth table for the logic gate. Look at Figure 2.4; there, we show the diagram that is used to represent the NAND gate in circuits and the truth table that the NAND gate obeys. The function that the NAND gate obeys is represented as an equation in Equation 2.4. The rule that this gate obeys can be stated as follows: When input X is logic High AND input Y is logic High, then the output from the NAND gate is logic Low. When any one of the two inputs or both the inputs are logic Low, then the output from the NAND gate is logic High.

$$F = \overline{(X \bullet Y)} = !(X \bullet Y) = \overline{X} + \overline{Y} \tag{2.4}$$

X	Y	F
0	0	1
0	1	1
1	0	1
1	1	0

Logic Diagram for NAND Gate.

Truth Table for NAND Gate.

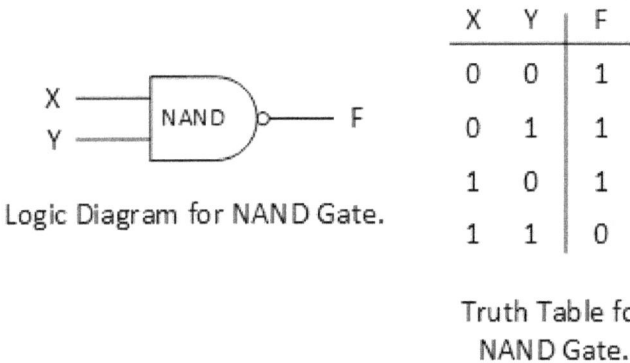

FIGURE 2.4. Logic Diagram and Truth Table for the NAND Gate.

We read Equation 2.4 as "F equals Not X and Y" or as "F equals X and Y bar" or as "F equals X bar or Y bar." This information in the truth table for this logic gate is the same as the logic equation. Let us see how this works. The first line in the truth table reads as: When input X is Low and input Y is Low, then the output F is High. We read the second line as: When input X is Low and input Y is High, then the output F is High. We read the third as: When input X is High and input Y is Low, then the output F is High. We read the last line as: When input X is High and input Y is High, then the output F is Low. Thus, both the truth table and the logic equation convey the same information. Only when both the inputs are High will the output be Low.

2.5. THE NOR GATE

The next logic gate that we will examine is a two-input NOR gate. To learn how to use the gate, we need to look at the function that the gate represents. The NOR gate can be represented in terms of a logic equation and the truth table for the logic gate. Look at Figure 2.5; there, we show the diagram that is used to represent the NOR gate in circuits and the truth table that the NOR gate obeys. The function that the NOR gate obeys is represented as an equation in Equation 2.5. The rule that this gate obeys can be stated as follows: When input X is logic Low and input Y is logic Low, then the output

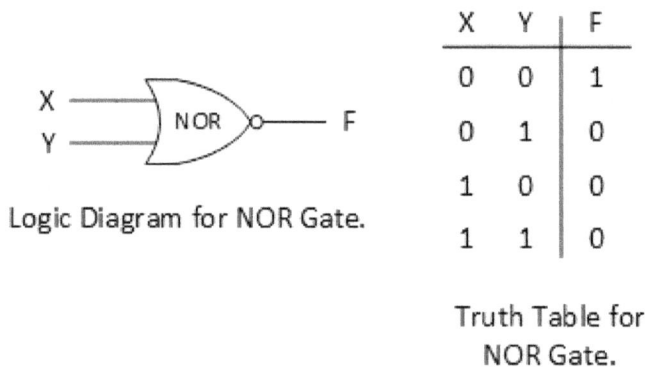

X	Y	F
0	0	1
0	1	0
1	0	0
1	1	0

Logic Diagram for NOR Gate.

Truth Table for NOR Gate.

FIGURE 2.5. Logic Diagram and Truth Table for the NOR Gate.

from the NOR gate is logic High. When either of the two inputs is logic High, then the output from the NOR gate is logic Low.

$$F = \overline{(X + Y)} = !(X + Y) = \overline{X} \bullet \overline{Y} \qquad (2.5)$$

We read Equation 2.5 as "F equals Not X or Y" or as "F equals X or Y bar" or as "F equals X bar and Y bar." This same information is present in the truth table for the logic gate. Let us see how this works. The first line in the truth table reads as: When input X is Low and input Y is Low, then the output F is High. We read the second line as: When input X is Low and input Y is High, then the output F is Low. We read the third as: When input X is High and input Y is Low, then the output F is Low. We read the last line as: When input X is High and input Y is High, then the output F is Low. Thus, both the truth table and the logic equation convey the same information. When both the inputs are Low, then the output will be High; when any one of the two inputs is High, then the output is going to be Low.

2.6.   THE XOR GATE

The next logic gate that we will examine is a two-input XOR gate. To learn how to use the gate, we need to look at the function that the gate represents. The XOR gate can be represented in terms of a logic equation and the truth table for the logic gate. Look at Figure 2.6; there, we show the diagram that is used to represent the NOR gate in circuits and the truth table that the XOR gate obeys. The function that the XOR gate obeys is represented as an equation in Equation 2.6. The rule that this gate obeys can be stated as follows: When the two inputs X,Y are both the same (either High or Low), then the output is Low. When the two inputs X,Y are complements of each other (one High and the other one Low), then the output is High.

$$F = X \oplus Y \qquad (2.6)$$

We read Equation 2.6 as "F equals X exclusive or Y." This same information is present in the truth table for the logic gate. Let us see how this works. The first line in the truth table reads as: When input X is Low and input Y

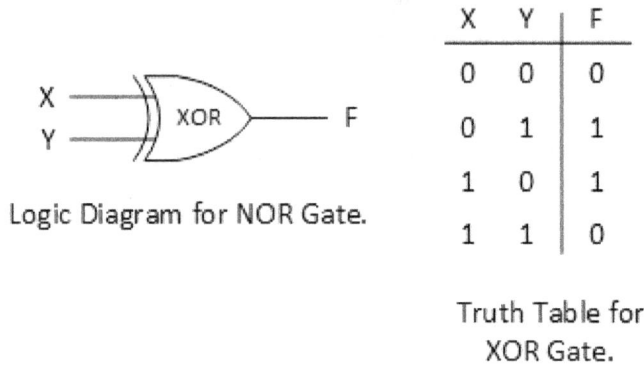

X	Y	F
0	0	0
0	1	1
1	0	1
1	1	0

Logic Diagram for NOR Gate.

Truth Table for XOR Gate.

FIGURE 2.6. Logic Diagram and Truth Table for the XOR Gate.

is Low, then the output F is Low. We read the second line as: When input X is Low and input Y is High, then the output F is High. We read the third as: When input X is High and input Y is Low, then the output F is High. We read the last line as: When input X is High and input Y is High, then the output F is Low. Thus, both the truth table and the logic equation convey the same information. When both the inputs are same, then the output will be Low; when the two inputs are complements of each other, then the output is going to be High.

2.7. THE XNOR GATE

The next logic gate that we will examine is a two-input XNOR gate. To learn how to use the gate, we need to look at the function that the gate represents. The XNOR gate can be represented in terms of a logic equation and the truth table for the logic gate. Look at Figure 2.7; there, we show the diagram that is used to represent the XNOR gate in circuits and the truth table that the XNOR gate obeys. The function that the XNOR gate obeys is represented as an equation in Equation 2.7. The rule that this gate obeys can be stated as follows: When the two inputs X,Y are both the same (either High or Low), then the output is High. When the two inputs X,Y

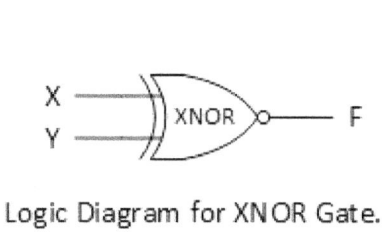

Logic Diagram for XNOR Gate.

X	Y	F
0	0	1
0	1	0
1	0	0
1	1	1

Truth Table for
XNOR Gate.

FIGURE 2.7. Logic Diagram and Truth Table for the XNOR Gate.

are complements of each other (one High and the other one Low), then the output is Low.

$$F = \overline{X \oplus Y} \qquad (2.7)$$

We read Equation 2.7 as "F equals X exclusive NOR Y." This same information is present in the truth table for the logic gate. Let us see how this works. The first line in the truth table reads as: When input X is Low and input Y is Low, then the output F is High. We read the second line as: When input X is Low and input Y is High, then the output F is Low. We read the third as: When input X is High and input Y is Low, then the output F is Low. We read the last line as: When input X is High and input Y is High, then the

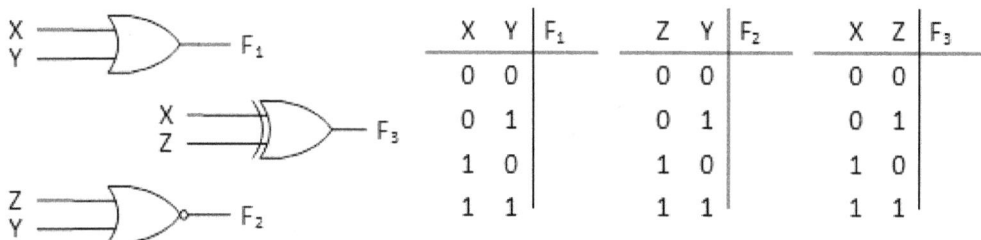

X	Y	F_1		Z	Y	F_2		X	Z	F_3
0	0			0	0			0	0	
0	1			0	1			0	1	
1	0			1	0			1	0	
1	1			1	1			1	1	

FIGURE 2.8. Logic Diagram for Review Question 2.1.

output F is High. Thus, both the truth table and the logic equation convey the same information. When both the inputs are the same, then the output will be High; when the two inputs are complements of each other, then the output is going to be Low. For this reason, this gate is sometimes referred to as the "Equivalence" gate.

Review Question for Sections 2.1 to 2.7

Question: Examine the logic diagrams given. For each diagram, develop a truth table for the output using the truth tables for the individual logic gates.

> **Answer:** The output F_1 is the output of an OR gate, so the truth table for the output F_1 is in order [0,1,1,1]. Similarly, the output F_2 is a NOR gate; this truth table will be [1,0,0,0]. Similarly, the output F_3 is a XOR gate; this truth table will be [0,1,1,0].

2.8.   USING THE LOGIC GATES

Now that we know the various different logic gates, we examine how we use these logic gates. The simplest way to use the logic gates is to connect a logic signal to the gate input terminal. The output of one logic gate can also be used as a logic signal, so this can be an input to another logic gate. When we interconnect these logic gates in this way, we build a logic function. The logic function that the interconnected logic gates build can be represented as either a logic equation or as a truth table. In this section, we examine both these different ways of representing a logic function.

2.8.1.   BUILDING THE LOGIC EQUATION

When we have several logic gates interconnected as shown in Figure 2.9, we need a way to tell someone what the logic function represents. We do this by a logic equation. As a simple example, we look at the logic function

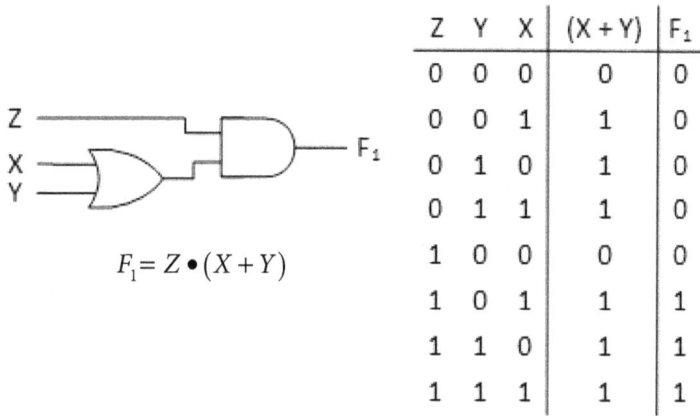

Z	Y	X	(X + Y)	F_1
0	0	0	0	0
0	0	1	1	0
0	1	0	1	0
0	1	1	1	0
1	0	0	0	0
1	0	1	1	1
1	1	0	1	1
1	1	1	1	1

$$F_1 = Z \bullet (X + Y)$$

FIGURE 2.9. A Simple logic Function.

represented in Figure 2.9. This logic function has three inputs and one output. To build the logic equation, we begin from the input side. The first gate that we see is the OR gate. The output of the OR gate can be written as $(X + Y)$. This output from the OR gate is one of the inputs to an AND gate; the other input to this AND gate is the logic variable Z. When the two inputs to the AND gate are combined, we get the output logic variable labeled as F_1. The combination of the AND gate can be written as $Z \bullet (X + Y)$. Since this is also the output, we have obtained the logic equation for the logic function.

2.8.2. BUILDING THE TRUTH TABLE

To represent the logic function as a truth table, we first write out all the possible combinations of all the input variables. This is done in Figure 2.9 for the three input variables X, Y, and Z. Next, we start building the truth table for the output for each gate. In the logic diagram, we see that the OR gate is the first gate on the input side. Combine the variables that are the input to the OR gate using the truth table of the OR gate. This is shown in the column headed (X + Y). Next, the output of the OR gate is combined with the input variable Z to give us the output from the logic function. To do this, we combine the logic values in column Z with the logic values in column

Z	Y	X	$\overline{(X+Y)}$	F_1
0	0	0	1	1
0	0	1	0	1
0	1	0	0	1
0	1	1	0	1
1	0	0	1	0
1	0	1	0	1
1	1	0	0	1
1	1	1	0	1

$$F_1 = \overline{Z \bullet (X+Y)}$$

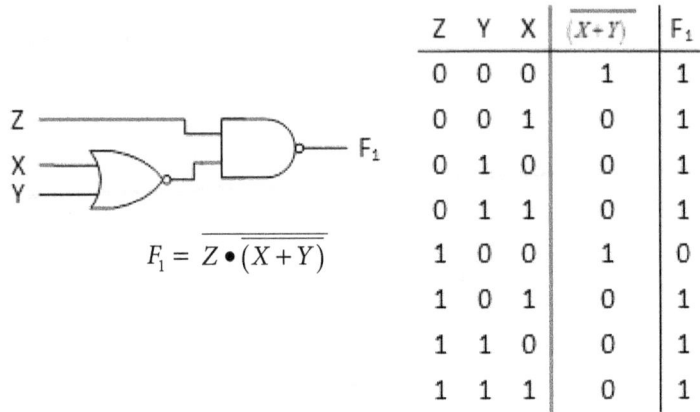

FIGURE 2.10. Determine the Logic Equation and build a Truth Table.

(X + Y) using the AND function to get the column F_1. This completes the truth table for the logic function.

Review Question and Answer: For the logic function in Figure 2.10, determine the logic equation and build the truth table.

2.9. READING DATA SHEETS

We have been using the logic values "0" and "1" to build truth tables for the various logic gates. These logic gates are packaged in a plastic casing that is invariably black with pins sticking out of the plastic casing. These pins represent the input to the gates and the output from the gates. There are no pictures or logic symbols on the outside of the plastic package, so how do we know what is inside the plastic package, and how are the pins connected to the logic gates? The answer is a "data sheet." Companies that make these logic gates also supply us with the data sheet for the logic gates that they make. The data sheet contains a lot of information, and knowing how to read the data sheet is essential. In this section, we indicate some of the important information that is on the data sheet and how a student can find it on the data sheet.

2.9.1. READING THE DATA SHEET FOR A LOGIC IC

The logic gates that we have just seen all come packaged in a plastic package. The outside of all these plastic packages look alike, so to distinguish one logic gate type from another we make use of data sheets. The data sheet for an AND gate IC is included in the appendix for you. There is a lot of information in the data sheet. For now, we only need some simple information from the data sheet.

2.9.1.1.　　　First, we need to know which gate and IC the data sheet represents. This is the top right hand side of the data sheet.

2.9.1.2.　　　Next, we need to know what each of the pins on the IC represent. Below the title of the data sheet, we see the pin diagram of the data sheet. The pin diagram tells us the function of each pin and what we should connect to the pin so that we can correctly use the IC. Notice that the pins are also numbered. Once you have identified pin number 1, then all the other pins are numbered in a counterclockwise direction. Pin number 1 is generally marked on the IC by either an indentation next to it or a small bump next to it.

2.9.1.3.　　　Third, toward the bottom of the page, we see the truth table and the logic diagram for the IC. Relate the information from the truth table to the pin diagram for the IC. In the truth table, we have the columns A and B. On the pin diagrams, we have labels 1A, 1B, etc. These labels imply that we have several different gates that perform similar functions and the pins 1A and 1B combine together to give output 1Y according to the logic equation described.

2.9.1.4.　　　On the next page, we are given a table of "Absolute Maximum Ratings." These represent the absolute maximum current and voltages that can be present in the IC for the IC to function as it is intended to. Be very careful of the voltage rating. Do not connect a voltage that is outside these boundaries on any pin on the IC. Connecting excessive voltage is about the only way that you can damage the IC, so please be careful.

2.9.1.5.　　　After the absolute maximum ratings, you see a table of "Recommended Operating Conditions." This tells us how we should operate the IC. This table also tells us what voltages to expect on the output when we expect a logic Low and a logic High output.

2.9.1.6.   The remaining information is about the timing and the package information. We will examine the timing information in one of the later chapters. The package information does not matter to us for now.

Review Question for Section 2.9. From the data sheet, identify the following:

a.   This Data sheet is for which IC?

b.   How many gates are in this one IC?

c.   What is the logic equation for each of the gates on the IC?

d.   What is the input voltage range for each of the input pins?

e.   What is the minimum voltage output for a logic High output?

a.   This Data Sheet is for the IC 74ALVC08, which is a quad two-input AND gate.

b.   There are four gates on this IC.

c.   The logic equation for each gate is either $Y = A \bullet B$ or $Y = \overline{(\overline{A} + \overline{B})}$.

d.   The input voltage range for each pin is −0.5 volts to 4.6 volts.

e.   There are several, depending on the supply voltage. For supply voltage in the range of 2.7 volts to 3.6 volts, the minimum high output voltage is 2.0 volts.

2.10. CHAPTER PROBLEMS

2.10.1. Draw the logic diagram, write the logic equation, and build a truth table for the AND logic gate.

2.10.2. Draw the logic diagram, write the logic equation, and build a truth table for the OR logic gate.

2.10.3. Draw the logic diagram, write the logic equation, and build a truth table for the NAND logic gate.

2.10.4. Draw the logic diagram, write the logic equation, and build a truth table for the NOR logic gate.

2.10.5. Draw the logic diagram, write the logic equation, and build a truth table for the XOR logic gate.

2.10.6. Draw the logic diagram, write the logic equation, and build a truth table for the NOT logic gate.

2.10.7. For the logic function $F_1 = x \bullet (y + \bar{z})$, build a truth table and draw the logic diagram.

2.10.8. For the logic function $F_2 = (x + z) + (y + \bar{z})$, build a truth table and draw the logic diagram.

2.10.9. For the logic function $F = \bar{x} \bullet (y + \bar{z})$, build a truth table and draw the logic diagram.

2.10.10. For the logic function $F = x \bullet y \bullet z + x \bullet \bar{z} + \bar{x} \bullet y + \bar{x} \bullet \bar{y} \bullet z$, build a truth table and draw the logic diagram.

3. BOOLEAN ALGEBRA AND WORKING WITH LOGIC EQUATIONS

3.0. INTRODUCTION TO LOGIC EQUATIONS AND MINIMIZATIONS

In this chapter, we begin with the representation of a function as a logic equation. When we write logic functions, we generally represent them as a logic equation in one of the two canonical forms. These two forms are known as the *Sum of Products* (SOP) form and the *Product of Sums* (POS) form. The two forms are distinctive in that they give us an indication of how the function is to be implemented. Implementing a logic function means we use the logic gates to build the function whose output is a logic signal that obeys the logic function. A logic function can be built using logic gates in many different ways. Some of these ways are more costly than others. We will always try to simplify a logic equation so that its cost to implement the logic function in one of the canonical forms is minimized. There are several different methods that you can use to minimize a logic function, and we will learn these as we progress. All the methods for logic function minimization are based on the simple laws and theorems of Boolean Algebra. In this chapter, we examine the laws and the theorems of Boolean Algebra.

3.1. LAWS, THEOREMS, AND AXIOMS OF BOOLEAN ALGEBRA

In the last chapter, we saw that the logic operations like the AND, OR, and the NOT can be represented by logic gates. Using these same logic operations, we can represent any and every logic function. Since we are using only the AND, OR, and the NOT logic operations, we can build any logic function using only the three gates AND, OR, and NOT. It is due to this close relationship between the logic gates and the laws of Boolean Algebra that we can use Boolean Algebra to manipulate logic functions and come up with equivalent functions. So a logic function is not unique; there can be many different logic functions that give us the same logic operation. What we look for is a logic function that is equivalent to the given function but uses fewer gates and/or inputs. We say that two functions are equivalent when the two of them have the same truth table.

3.1.1 THE AXIOMS OF BOOLEAN ALGEBRA

Boolean Algebra consists of a set of elements (which we call logic variables) together with two binary operations (the AND and the OR) and one unary operation (the NOT). The three operators operate on the variables as follows.

a. All the variables can take on two values, δ and θ such that $\delta \neq \theta$. In our binary system, the two values that we use are 1 and 0.

b. Closure of the set: When logic variables are combined together using the logic operations, the result of the combination will be one of the two values. Equation 3.1 says that when we apply the logic operation of the AND or the OR on any two binary values, the result will be a value that is one of the two binary values.

$$\alpha \bullet \beta = \text{either } \delta \text{ or } \theta \qquad \text{(The AND Operation)}$$
$$\alpha + \beta = \text{either } \delta \text{ or } \theta \qquad \text{(The OR Operation)} \qquad (3.1)$$

c. Commutative Laws: The order of the variables in any combination is immaterial. It does not matter which input is considered first. When the two inputs are combined, the result is the same no matter which order you look at the inputs. This is shown in Equation 3.2.

$$\alpha \bullet \beta = \beta \bullet \alpha \qquad \text{(The AND Operation)}$$
$$\alpha + \beta = \beta + \alpha \qquad \text{(The OR Operation)}$$

(3.2)

d. Associative Law: When you have three inputs, then you can take any two inputs first and then combine the result with the third input. This is shown in Equation 3.3.

$$\alpha \bullet \beta \bullet \gamma = (\alpha \bullet \beta) \bullet \gamma = \alpha \bullet (\beta \bullet \gamma)$$
$$\alpha + \beta + \gamma = (\alpha + \beta) + \gamma = \alpha + (\beta + \gamma)$$

(3.3)

e. Identities: Each of the two operators has an identity element such that when the operation of any value is applied with the identity element, the result is the same value as that of the element. This is shown in Equation 3.4a for the AND operation and in 3.4b for the OR operation.

AND operation	$\alpha \bullet 1 = \alpha$	(3.4a)
OR operation	$\alpha + 0 = \alpha$	(3.4b)

Equation 3.4a says that no matter what the value of variable α is, after the AND operation with 1, the result will always be the value that the variable α had. If α is zero, then (1 AND 0) = 0, the same value as α on the other hand, if α is zero, then (1 AND 1) = 1, the same value as α. Similarly, Equation 3.4b says the same for the OR operation. For the OR operation, the identity value is the value of 0.

f. Distributive Law: This law tells us how the two operations interact with each other. This is shown in Equation 3.5.

$$(\alpha \bullet \beta) + \gamma = (\alpha + \gamma) \bullet (\beta + \gamma)$$
$$(\alpha + \beta) \bullet \gamma = (\alpha \bullet \gamma) + (\beta \bullet \gamma)$$

(3.5)

g. Complement: For every value, there exists a complement of that value. When we use the logic operations with the complements, we get a known result for both the AND and the OR operations. Operations with complements are shown in Equation 3.6. The AND operation with its complement always gives a zero result, while the OR operation with the complement always gives a 1 result.

$$\alpha \bullet \bar{\alpha} = 0$$
$$\alpha + \bar{\alpha} = 1$$

(3.6)

It is easy to show that the two binary values of {0, 1} and the logical operations of AND, OR, and NOT satisfy all the above axioms. The simplest way is by exhaustive search. It only means that we try out every possible combination of values and verify the result. To do this, you substitute the numerical values of 0 and 1 for the inputs α and β Next, replace the + with the OR operation, the $\bullet$ with the AND operation, and the over bar for the NOT operation, and you will get the expected results. For example, let us examine the identity element from Equations 3.4a and 3.4b in Equation 3.7.

$$0 + 1 = 1 \quad \text{and} \quad 1 + 1 = 1 \quad \text{Identity element for the OR operation.}$$
$$0 \bullet 0 = 0 \quad \text{and} \quad 1 \bullet 0 = 0 \quad \text{Identity element for the AND operation.}$$

(3.7)

So we see that when the OR operation is used, 1 is the identity element. This can be verified from the truth table for the OR gate from Figure 2.2 and Equation 2.2. When the AND operation is used, 0 is the identity element. This is verified from the truth table for the AND gate from Figure 2.1 and Equation 2.1. Using this same procedure, all the axioms can be verified.

Review Questions for Section 3.1.1

Question: Verify the validity of the Closure Axiom.

    Answer: The truth table for the OR gate shows us all the possible combinations for two variables with the OR operation. When we look at the output column in the truth table, we see that all the values on the output column are either 0 or 1. This proves that the OR operation gives us as a result one of two logic values; thus, the closure for the OR function is verified. Similarly, writing the truth table for the AND function will prove the closure for the AND function.

Question: Verify the Commutative Law for Boolean Algebra.

    Answer: To verify the Commutative Law for the OR operation, we write the result of the OR operation for the two values, as shown in Equation 3.8.

$$\begin{aligned} 0+1 &= 1+0 \quad \text{Statement of Commutative Law} \\ 1 &= 1 \quad \text{Result of evaluating the two sides.} \end{aligned} \tag{3.8}$$

To verify the Commutative Law for the AND operation, we write the result of the AND operation for the two values, as shown in Equation 3.9.

$$\begin{aligned} 0 \bullet 1 &= 1 \bullet 0 \quad \text{Statement of Commutative Law} \\ 0 &= 0 \quad \text{Result of evaluating the two sides.} \end{aligned} \tag{3.9}$$

3.1.2. LAWS OF BOOLEAN ALGEBRA AND OPERATIONS WITH LOGIC VALUES

Here, we list and discuss the frequently used laws and theorems of Boolean Algebra. Some of these we already saw in the previous section, but they are presented here in a more general manner and in the way that we will be using them most often. We also list them in two separate columns: one column for the OR operation and the other column for the AND operation.

a. Variables combining with the numerical values of 0 and 1.

$$X+0=X \qquad\qquad X \bullet 0 = 0$$
$$X+1=1 \qquad\qquad X \bullet 1 = X \tag{3.10}$$

b. The Idempotent Theorem

$$X+X=X \qquad\qquad X \bullet X = X \tag{3.11}$$

Many students have difficulty with this theorem when the value of X is 1. In this case, the equation for the OR identity is "1 + 1 = 1." This confusion will continue if you read the "+" symbol as "plus." This confusion will go away if you read the "+" symbol as OR. From now on, we will always read the equation as "X OR X equals X."

c. Involution Theorem: The complement of the complement of a logic variable is the original variable itself. The two complements cancel each other.

$$\overline{\left(\overline{X} \right)} = X \tag{3.12}$$

d. Complements: A variable when combined with its complement gives you the identity value under the operation.

$$X+\overline{X}=1 \qquad\qquad X \bullet \overline{X} = 0 \tag{3.13}$$

e. Commutative Law: The order in which the variables enter the operation does not matter.

$$X+Y=Y+X \qquad\qquad X \bullet Y = Y \bullet X \tag{3.14}$$

f.   Distributive Law: This law is a little difficult to see if you read the "+" as plus. Again, it will be easy to see if you replace the plus with OR.

$$\left(X+Y\right)\bullet Z=\left(X\bullet Z\right)+\left(Y\bullet Z\right)\ \ \left(X\bullet Y\right)+Z=\left(X+Z\right)\bullet\left(Y+Z\right)$$

(3.15)

g.   Simplification Theorems: There are a few logic expressions that aid in simplifying (or writing a logic function with a cheaper cost).

$$X\bullet Y+X\bullet\bar{Y}=X \qquad\qquad \left(X+Y\right)\bullet\left(X+\bar{Y}\right)=X \quad (3.16a)$$

$$X+X\bullet Y=X \qquad\qquad\qquad X\bullet\left(X+Y\right)=X \qquad\quad (3.16b)$$

$$\left(X+\bar{Y}\right)\bullet Y=X\bullet Y \qquad\qquad \left(X\bullet\bar{Y}\right)+Y=X+Y \qquad (3.16c)$$

$$X\bullet Y+X\bullet\bar{Y}=X \qquad\qquad \left(X+Y\right)\bullet\left(X+\bar{Y}\right)=X \quad (3.16d)$$

These are the simple laws of Boolean Algebra. Using these laws allows us to re-write the Boolean expressions in alternate formats. Next, we examine how we will use these laws of Boolean Algebra to simplify (write in an alternate format) a logic expression.

Example: You are given the logic expression in Equation 3.17. Using the laws of Boolean Algebra, simplify the logic expression so that it can be represented using fewer gates and/or fewer inputs. Figure 3.1 shows us how we would build the function as it is given. In Figure 3.1, we see that we need five gates (four AND gates and one OR gate) and sixteen inputs to the gates (three inputs to each of the four AND gates for twelve and then four inputs to the OR gate for a total of sixteen) to build the function. (We do not count inverters when we are counting the number of gates.)

$$f=\bar{x}\bullet y\bullet z+x\bullet\bar{y}\bullet z+x\bullet y\bullet\bar{z}+x\bullet y\bullet z \qquad (3.17)$$

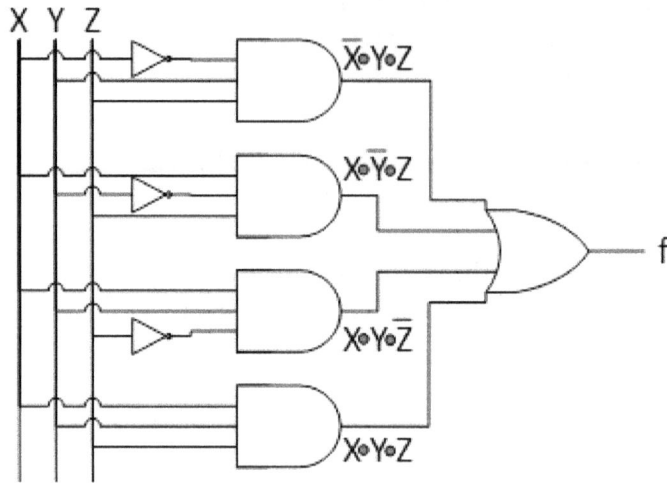

FIGURE 3.1. Building a logic function.

To build the function that is of cheaper cost, we will follow the sequence shown in Equation 3.18 where we first use the *Idempotent Theorem* and write the term $x \bullet y \bullet z = x \bullet y \bullet z + x \bullet y \bullet z$. Next, we use one of the two terms on the right to combine with the term $\bar{x} \bullet y \bullet z$, first using the *Simplification Theorem d* to factor the common terms out and then using the *Complements Theorem* to write $(x + \bar{x}) = 1$. Finally, using the *Variables Combining with Numerical Values Theorem*, we get the last expression in Equation 3.18.

$$f = \bar{x} \bullet y \bullet z + x \bullet \bar{y} \bullet z + x \bullet y \bullet \bar{z} + x \bullet y \bullet z + x \bullet y \bullet z \qquad \text{Idempotent Theorem}$$
$$f = y \bullet z \bullet (\bar{x} + x) + x \bullet \bar{y} \bullet z + x \bullet y \bullet \bar{z} + x \bullet yy \bullet z \qquad \text{Simplification Theorem d}$$
$$f = y \bullet z \bullet 1 + x \bullet \bar{y} \bullet z + x \bullet y \bullet \bar{z} + x \bullet y \bullet z \qquad \text{Complements Theorem}$$
$$f = y \bullet z + x \bullet \bar{y} \bullet z + x \bullet y \bullet \bar{z} + x \bullet y \bullet z \qquad \text{Combining with Numerical Values}$$

(3.18)

If we now compare the last expression with the original expression, we know that the two of them perform the same function, but the last expression in Equation 3.18 has one less input (in the first term). We say that the expression in Equation 3.18 has lower cost than the original expression in

Equation 3.17 as it requires fewer inputs and/or fewer gates. We can continue with this same procedure once more, as shown in Equation 3.19, to get an expression that performs the same logic expression but is cheaper in cost than the original expression.

$$f = y \bullet z + x \bullet \overline{y} \bullet z + x \bullet y \bullet \overline{z} + x \bullet y \bullet z + x \bullet y \bullet z \qquad \text{Idempotent Theorem}$$
$$f = y \bullet z \bullet + x \bullet z \bullet (\overline{y} + y) + x \bullet y \bullet \overline{z} + x \bullet y \bullet z \qquad \text{Simplification Theorem d}$$
$$f = y \bullet z + x \bullet z \bullet 1 + x \bullet y \bullet \overline{z} + x \bullet y \bullet z \qquad \text{Complements Theorem}$$
$$f = y \bullet z + x \bullet z + x \bullet y \bullet \overline{z} + x \bullet y \bullet z \qquad \text{Combining with Numerical Values}$$

$$(3.19)$$

The last expression in Equation 3.19 is cheaper to build than the expression in Equation 3.17 or Equation 3.18, yet all three of them perform the same logic operation. We can repeat the same process one more time on the last expression in Equation 3.19 to get a new expression, shown in Equation 3.20.

$$f = y \bullet z + x \bullet z + x \bullet y \bullet \overline{z} + x \bullet y \bullet z \qquad \text{Idempotent Theorem}$$
$$f = y \bullet z \bullet + x \bullet z + x \bullet y \bullet (\overline{z} + z) \qquad \text{Simplification Theorem d}$$
$$f = y \bullet z + x \bullet z + x \bullet y \bullet 1 \qquad \text{Complements Theorem}$$
$$f = y \bullet z + x \bullet z + x \bullet y \qquad \text{Combining with Numerical Values}$$

$$(3.20)$$

When we examine Equation 3.20, we see that the last equation in Equation 3.20 performs the same function as the original logic equation given in Equation 3.17 but uses fewer gates. The function in Equation 3.20 uses four gates (three AND gates and one OR gate) and only nine inputs (two inputs to each of the three AND gates for six and then three inputs to the OR gate for a total of nine). To build this function, we need only four gates and nine inputs in this form; in the original form, we required five gates and sixteen inputs. This is a great reduction in cost. We see the logic diagram to build the function given in Equation 3.20 is given in Figure 3.2, and it looks a lot simpler.

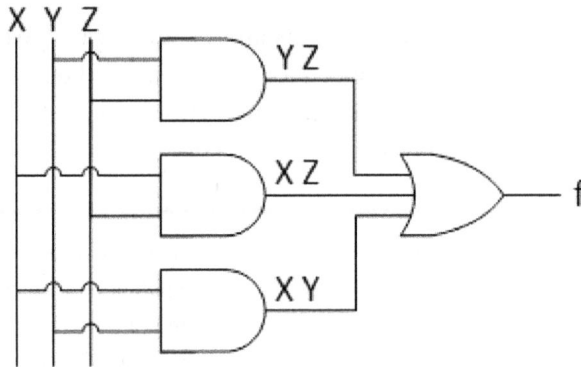

FIGURE 3.2. Building a cheaper login function.

Review Question for Section 3.1.2

Question: Verify the validity of all the four Simplification Theorems.
Answer: To prove theorem "a," we look at Equation 3.21.

$$X \bullet Y + X \bullet \bar{Y} = X \bullet (Y + \bar{Y}) = X \bullet 1 = X$$

$$
\begin{aligned}
&(X + Y) \bullet (X + \bar{Y}) \\
&X \bullet X + X \bullet Y + X \bullet \bar{Y} + Y \bullet \bar{Y} \\
&X + X \bullet Y + X \bullet \bar{Y} + 0 \\
&X
\end{aligned}
$$

$$(3.21)$$

To prove theorem "b," we look at Equation 3.22.

$$X + X \bullet Y = X \bullet (1 + Y) = X \bullet 1 = X$$

$$
\begin{aligned}
&X \bullet (X + Y) \\
&X \bullet X + X \bullet Y \\
&X \bullet (1 + Y) \\
&X
\end{aligned}
$$

$$(3.22)$$

To prove theorem "c," we look at Equation 3.23.

$$
\begin{aligned}
&(X + \bar{Y}) \bullet Y \\
&X \bullet Y + Y \bullet \bar{Y} \\
&X \bullet Y + 0 \\
&X \bullet Y
\end{aligned}
$$

$$
\begin{aligned}
&X \bullet \bar{Y} + Y \\
&X \bullet \bar{Y} + (X + \bar{X}) \bullet Y \\
&X \bullet \bar{Y} + X \bullet Y + \bar{X} \bullet Y \\
&X \bullet (\bar{Y} + Y) + (X + \bar{X}) \bullet Y \\
&X + Y
\end{aligned}
$$

$$(3.23)$$

To prove theorem "d," we look at Equation 3.24.

$$
\begin{array}{ll}
X \bullet Y + X \bullet \bar{Y} & \left(X + Y\right) \bullet \left(X + \bar{Y}\right) \\
X \bullet \left(Y + \bar{Y}\right) & X \bullet X + X \bullet Y + X \bullet \bar{Y} + Y \bullet \bar{Y} \\
X \bullet 1 & X + X \bullet Y + X \bullet \bar{Y} + 0 \\
X & X \bullet \left(1 + Y\right) + X \bullet \left(1 + Y\right) \\
& X
\end{array}
\tag{3.24}
$$

3.1.3. DEMORGAN'S THEOREM

This law gives us a procedure to complement a logic function or a logic expression. The procedure is simple: when we apply the DeMorgan's Theorem to a logic operation, the result is the complement of the original logic function. To apply the DeMorgan's Theorem, we complement all the variables, we replace all the 1's by 0's and all the 0's by 1's, and finally, we replace all the AND operations with OR operations and the OR operations with AND operations. Examine Equation 3.25 to see how this comes about.

$$
\overline{\left(X + Y\right)} = \bar{X} \bullet \bar{Y} \qquad\qquad \overline{\left(X \bullet Y\right)} = \bar{X} + \bar{Y}
\tag{3.25}
$$

In Equation 3.25, we have the expression (X + Y) on the left hand side. We want to determine the complement of this expression. To do this, we place a complement over the expression and then apply the DeMorgan's Theorem. The theorem requires us to complement each variable and then to complement the operation. To do this, we have replaced all the variables with their complements and have changed the OR operation to the AND operation.

In the second equation in Equation 3.25, the logic function that we want to complement is the logic function $\left(X \bullet Y\right)$. To complement this function, we apply DeMorgan's Theorem. The theorem requires us to complement each variable and then to change the AND operation to the OR operation. To verify the validity of DeMorgan's Theorem, let us use the truth table method to verify that the left hand side of a logic expression is equal to the right hand side. This is given in Table 3.1 below.

X	Y	$\bar{X}$	$\bar{Y}$	$\overline{(X+Y)}$	$\bar{X} \bullet \bar{Y}$
0	0	1	1	1	1
0	1	1	0	0	0
1	0	0	1	0	0
1	1	0	0	0	0

TABLE 3.1A DEMORGAN'S THEOREM OR OPERATION

X	Y	$\bar{X}$	$\bar{Y}$	$\overline{(X \bullet Y)}$	$\bar{X} + \bar{Y}$
0	0	1	1	1	1
0	1	1	0	1	1
1	0	0	1	1	1
1	1	0	0	0	0

TABLE 3.1B DEMORGAN'S THEOREM AND OPERATION

DeMorgan's Theorem also brings out an interesting relation between the NAND logic and the NOR logic. The equation on the left in Equation 3.25 says that the NOR function is the same as the AND function if we complement the variable before applying the AND operation. The equation on the right in Equation 3.25 says that the NAND operation is the same as the OR function if we complement the variable before applying the OR operation.

A very important corollary of this theorem is given in Equation 3.26.

$$\overline{(X+Y)} \neq \bar{X} + \bar{Y} \qquad \overline{(X \bullet Y)} \neq \bar{X} \bullet \bar{Y} \qquad (3.26)$$

The two expressions in Equation 3.26 are not equal to each other. This is due to the time when the complement operation is performed. In both the equations on the left hand side, the OR (AND) operation is performed first and then the result is complemented. In both the equations on the right hand side, the complement operation is first performed on the variables and then the OR(AND) operation is performed. These are not equal.

Let us use the DeMorgan's Theorem to determine the complement of the logic function in Equation 3.17. This is done in Equation 3.27.

$$\bar{f} = \overline{\left(\bar{x} \bullet y \bullet z + x \bullet \bar{y} \bullet z + x \bullet y \bullet \bar{z} + x \bullet y \bullet z \right)}$$

$$\bar{f} = \overline{\left(\bar{x} \bullet y \bullet z \right)} \bullet \overline{\left(x \bullet \bar{y} \bullet z \right)} \bullet \overline{\left(x \bullet y \bullet \bar{z} \right)} \bullet \overline{\left(x \bullet y \bullet z \right)}$$

$$\bar{f} = \left(\bar{\bar{x}} + \bar{y} + \bar{z} \right) \bullet \left(\bar{x} + \bar{\bar{y}} + \bar{z} \right) \bullet \left(\bar{x} + \bar{y} + \bar{\bar{z}} \right) \bullet \left(\bar{x} + \bar{y} + \bar{z} \right) \qquad (3.27)$$

$$\bar{f} = \left(x + \bar{y} + \bar{z} \right) \bullet \left(\bar{x} + y + \bar{z} \right) \bullet \left(\bar{x} + \bar{y} + z \right) \bullet \left(\bar{x} + \bar{y} + \bar{z} \right)$$

In the second line, we have complemented each entire group of variables and also the logic operation that connects the different groups together. It is a good idea to put parentheses around each group of variables so you know what has to stay together and what is not together. We repeat the same procedure in the third line in Equation 3.27. In this line, we apply the DeMorgan's Theorem to each group within the function. Finally, we apply the Involution Theorem to clean up and get the final expression in the last line in Equation 3.27. Notice that following the procedure step by step, the application of the DeMorgan's Theorem is very simple and gives us the complement of the entire logic function.

Review Question for Section 3.1.3

Question: Apply DeMorgan's Theorem to determine the complement of the given logic function in Equation 3.28.

$$f = w \bullet \bar{x} \bullet \bar{y} + \bar{w} \bullet \bar{x} \bullet \bar{y} + w \bullet \bar{x} \bullet y + \bar{w} \bullet x \bullet \bar{y} \qquad (3.28)$$

Answer: The result after going through the steps of the DeMorgan's Theorem are given in Equation 3.29.

$$\bar{f} = \overline{\left(w \bullet \bar{x} \bullet \bar{y} + \bar{w} \bullet \bar{x} \bullet \bar{y} + w \bullet \bar{x} \bullet y + \bar{w} \bullet x \bullet \bar{y} \right)}$$

$$\bar{f} = \overline{\left(w \bullet \bar{x} \bullet \bar{y} \right)} \bullet \overline{\left(\bar{w} \bullet \bar{x} \bullet \bar{y} \right)} \bullet \overline{\left(w \bullet \bar{x} \bullet y \right)} \bullet \overline{\left(\bar{w} \bullet x \bullet \bar{y} \right)}$$

$$\bar{f} = \left(\bar{w} + \bar{\bar{x}} + \bar{\bar{y}} \right) \bullet \left(\bar{\bar{w}} + \bar{\bar{x}} + \bar{\bar{y}} \right) \bullet \left(\bar{w} + \bar{\bar{x}} + \bar{y} \right) \bullet \left(\bar{\bar{w}} + \bar{x} + \bar{\bar{y}} \right) \qquad (3.29)$$

$$\bar{f} = \left(\bar{w} + x + y \right) \bullet \left(w + x + y \right) \bullet \left(\bar{w} + x + \bar{y} \right) \bullet \left(w + \bar{x} + y \right)$$

3.2. WHY MINIMIZE LOGIC EQUATIONS?

In the previous section, we learned the first method that is used for minimizing logic functions. In the chapters that follow, we will examine and learn at least two other methods. The methods that we will learn all minimize the logic function assuming a two-level implementation. In a two-level minimization, the first level is generally the AND gates and the second level is the OR gate. This will give us the Sum of Products (SOP) expression. Earlier, we saw that the AND and the OR functions have laws and theorems that are very similar to each other. As a result of this, the same procedure can be used to get a minimum expression that uses the OR gates at the first level and the AND gates at the second level. So the fundamental reason for minimizing a logic function is so that we use fewer gates and inputs to build the logic function, as we saw earlier in Figures 3.1 and 3.2. We do not minimize the logic function to use one type of logic gate or the other type. All the logic gates have the same cost. This leads to what we call the "Time and Space" tradeoff.

3.2.1. TIME AND SPACE TRADEOFF

There are essentially three variables that we have to consider when we are building a logic function. The first variable is the number of logic gates that are used to build a logic function. Each logic gate takes up physical space either on a printed circuit board or inside an integrated circuit. Since space on a printed circuit board or in an IC is limited, we can build more functions in the same physical space if we use fewer gates to build the logic function. There is also a strong correlation between the complexity of design and manufacture and the number of gates required in the design. The design that is simplest to manufacture is usually the one with the fewest gates.

The second variable that we examine is the number of inputs. Another impact of space is the number of inputs to each logic gate. The inputs are often referred to as literals. While we can have gates with two, three, or four literals, gates with more literals—say, eight or ten—are very rare, but they are available and can be used. In a logic circuit, the literals have to be routed from one place to another (from the source to the destination). This also increases complexity and takes up space, so having fewer literals leads to a simpler design that can be manufactured easily.

The third variable that we consider is the time it takes for the signal to go from the input in a logic function to the output of the logic function. To understand the impact of time, we must first see where time enters the picture. We all know that the electric current travels very fast at the speed of light. Fast this may be, but it does take some time. Also, as the electrical signal travels through a logic gate, it takes some extra time to travel through the gate due to the properties of the electronics within the gate. As a result, if a signal has to travel through many levels of logic gates, the signal will take longer to reach its final destination. We are talking about time that is of the order of $5*10^{-9}$ seconds. This is indeed very fast, but if we did not pay a lot of attention to the time it takes a signal to go from its source to its destination, our computers would not be able to run as fast as they do now. We all know how frustrating it is for the computer screen to load up with a web page. In general, the fewer logic levels the signal has to go through, the less time it will take to reach its destination.

Unfortunately, it is not possible to minimize over all the three different factors—gate count, the number of literals, and the number of levels—so in our study here, we will fix the number of logic levels to two and then minimize the logic function over the number of gates used and the number of literals required to build the function. One example will make this concept of tradeoff between the three different types of costs to build a logic function.

Example: To examine the effects of tradeoff, consider the logic function given in Equation 3.30.

$$f = \overline{w} \bullet \overline{x} \bullet y + w \bullet x \bullet \overline{y} + w \bullet \overline{x} \bullet y + \overline{w} \bullet x \bullet y \qquad (3.30)$$

Using brute force and taking the function as it is written in Equation 3.30, we could implement the function as shown in Figure 3.3. In the diagram, we

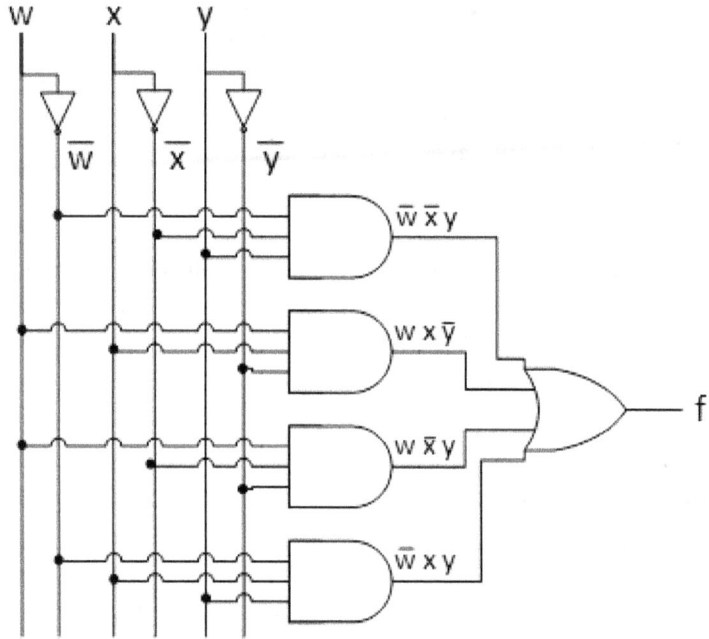

FIGURE 3.3. Implementing a logic function.

see that the function needs four three-input AND gates, three NOT gates, and one four-input OR gate. This implementation is known as two-level implementation. The first level is the level of the AND gates and the second level is the level of the OR gate.

If we use our theorems and laws of Boolean Algebra on this function, we can write this function as shown in Equation 3.31.

$$f = \overline{w} \bullet \overline{x} \bullet y + w \bullet x \bullet \overline{y} + w \bullet \overline{x} \bullet y + \overline{w} \bullet x \bullet y$$
$$f = \left(\overline{w} + w \right) \bullet \overline{x} \bullet y + w \bullet x \bullet \overline{y} + \left(\overline{x} + x \right) \bullet \overline{w} \bullet y \qquad (3.31)$$
$$f = \overline{x} \bullet y + w \bullet x \bullet \overline{y} + \overline{w} \bullet y$$

Implementing the function as shown in Equation 3.31 is a lot simpler than implementing it as shown in Figure 3.3. In arriving at this implementation, we have tried to keep the delay to two gate levels. We have reduced the number of gates and the number of literals. If we were allowed to implement the function with more than two gate levels, we can write the function as shown in Equation 3.32.

$$f = \overline{w} \bullet \overline{x} \bullet y + w \bullet x \bullet \overline{y} + w \bullet \overline{x} \bullet y + \overline{w} \bullet x \bullet y$$

$$f = \left(\overline{w} + w \right) \bullet \overline{x} \bullet y + w \bullet x \bullet \overline{y} + \left(\overline{x} + x \right) \bullet \overline{w} \bullet y$$

$$f = \overline{x} \bullet y + w \bullet x \bullet \overline{y} + \overline{w} \bullet y$$

$$f = \left(\overline{x} + \overline{w} \right) \bullet y + \left(w \bullet x \right) \bullet \overline{y} \qquad\qquad (3.32)$$

$$f = \left(x \bullet \overline{w} \right) \bullet y + \left(w \bullet x \right) \bullet \overline{y}$$

$$f = \left(w \bullet x \right) \oplus y$$

This last implementation uses the XOR function to build the same function. In this implementation, we first use one AND gate to combine the w and the x literals. This is one logic level. After the AND function, we have to build the XOR function using the two-level AND and OR gates. So this time, the longest path is through three gates but the number of literals has been reduced. You could say that the last equation in Equation 3.32 is only two levels and it uses only two gates. That would be true the way the logic equation is written, but that implies that we will go and get an XOR gate. An XOR gate is a complex gate and it is not as fast as the simple AND and the OR gates. So, you see there are many different ways to build a logic function, and the one that you choose to use has different tradeoffs. Since we are just starting out our study, we will concentrate mostly on two-level logic minimization so we can build the logic function using AND and OR gates or their equivalents.

3.2.2. WRITING LOGIC EQUATIONS IN CANONICAL FORMS

In the previous section, we decided that we will minimize logic functions so that they can be implemented using two-level logic. The two levels are always either AND on the first level and OR on the second level, which gives us the "Sum of Products" (SOP) canonical form, or the OR on the first level and AND on the second level, which gives us the "Product of Sums" (POS) canonical form. Both these forms are equal to each other, and the same logic function can be implemented using either form. Here, we will see how we can get the expressions for the two canonical forms from the truth table of the logic function.

The truth table of the function tells us when the output from the function will be logic High and under what condition the output from the logic function will be logic Low. The logic equation, on the other hand, gives us the combinations of the implants that gives us the logic High output or the logic Low output. Thus, the truth table and the logic equation are closely related to each other. To write either the SOP or the POS expressions, we will always begin with the truth table of the function.

Since we are going to talk about rows of the truth table, it would be nice to be able to identify the rows by a name rather than the logic expression for the row of the truth table. We do this in two different ways. To help us to understand how we name these rows, look at Figure 3.4.

The Min term column in Figure 3.4 tells us how to identify each row in the truth table using the Min terms. A Min term is typically a term that combines all the literals in the truth table as they appear in the truth table using the AND operation. Each literal will appear either in

W	X	Y	Min Terms	Max Terms
0	0	0	$\overline{w} \bullet \overline{x} \bullet \overline{y} \Rightarrow m_0$	$w + x + y \Rightarrow M_0$
0	0	1	$\overline{w} \bullet \overline{x} \bullet y \Rightarrow m_1$	$w + x + \overline{y} \Rightarrow M_1$
0	1	0	$\overline{w} \bullet x \bullet \overline{y} \Rightarrow m_2$	$w + \overline{x} + y \Rightarrow M_2$
0	1	1	$\overline{w} \bullet x \bullet y \Rightarrow m_3$	$w + \overline{x} + \overline{y} \Rightarrow M_3$
1	0	0	$w \bullet \overline{x} \bullet \overline{y} \Rightarrow m_4$	$\overline{w} + x + y \Rightarrow M_4$
1	0	1	$w \bullet \overline{x} \bullet y \Rightarrow m_5$	$\overline{w} + x + \overline{y} \Rightarrow M_5$
1	1	0	$w \bullet x \bullet \overline{y} \Rightarrow m_6$	$\overline{w} + \overline{x} + y \Rightarrow M_6$
1	1	1	$w \bullet x \bullet y \Rightarrow m_7$	$\overline{w} + \overline{x} + \overline{y} \Rightarrow M_7$

FIGURE 3.4. Identifying Min terms and Max terms.

its true form or in its complemented form, but not both. All the literals in the truth table must appear in the Min term. The literals in the Min term are combined using the AND operation. A number identifies the Min terms. The number is the decimal equivalent of the binary representation in the truth table. So Min term m_7 is identified as shown in Equation 3.33a.

$$m_7 = 111 = w \bullet x \bullet y \qquad (3.33a)$$

The Max term column in Figure 3.4 tells us how to identify each row in the truth table using the Max terms. A Max term is typically a term that combines all the literals in the truth table not as they appear, but after they are complemented using the OR operation. Each literal will appear either in its true form or in its complemented form, but not both. All the literals in the truth table must appear in the Max term. The literals in the Max term are combined using the OR operation. A number identifies the Max terms. The number is the complement of the decimal equivalent of the binary representation in the truth table. So Max term M_6 is identified as shown in Equation 3.33b.

$$M_6 = 001 = \bar{w} + \bar{x} + y \qquad (3.33b)$$

From now on, we will use either Min terms or Max terms to identify any logic function.

3.2.2.1.   THE SUM OF PRODUCTS FORM

The Sum of Products form gets its name from the way the logic function is written. The SOP form uses the Min terms to identify the function. Since each Min term is the result of ANDing the literals, this is identified as if we are multiplying the literals together, hence the term Products. Then all the Min terms are combined using the OR operation, hence the term Sum. To write the logic expression in the SOP form, we only look at those rows of the truth table that have a logic High output. We ignore the rows with the logic Low output. We identify these rows as the Min terms and add the Min

W	X	Y	Function 1	Function 2	Min Term	Max Term
0	0	0	0	1	m_0	M_0
0	0	1	0	0	m_1	M_1
0	1	0	1	0	m_2	M_2
0	1	1	1	1	m_3	M_3
1	0	0	1	0	m_4	M_4
1	0	1	1	1	m_5	M_5
1	1	0	0	1	m_6	M_6
1	1	1	0	0	m_7	M_7

FIGURE 3.5. Writing a Logic function from a Truth Table using the SOP format or the POS format.

terms to get the SOP form of the logic function. The examples in Figure 3.5 will make this clear.

To write function 1, we see that we have only four entries that are logic High. They correspond to the Min terms m_2, m_3, m_4, and m_5. We will build this function by combining these four rows (Min terms) by an OR gate. So to write the function F_1, we would "Sum" the four Min terms that have the logic High output, as shown in Equation 3.34.

$$F_1 = \sum \left(m_2, m_3, m_4, m_5 \right)$$
$$F_1 = \overline{w} \bullet x \bullet \overline{y} + \overline{w} \bullet x \bullet y + w \bullet \overline{x} \bullet \overline{y} + w \bullet \overline{x} \bullet y \qquad (3.34)$$

To write function 2, we see that we have only four entries that are logic High. They correspond to the Min terms m_0, m_3, m_5, and m_6. We will build this function by combining these four rows (Min terms) by an OR gate. So to write the function F_2, we would "Sum" the four Min terms that have the logic High output, as shown in Equation 3.35.

$$F_2 = \sum \left(m_0, m_3, m_5, m_6 \right)$$
$$F_2 = \overline{w} \bullet \overline{x} \bullet \overline{y} + \overline{w} \bullet x \bullet y + w \bullet \overline{x} \bullet y + w \bullet x \bullet \overline{y} \qquad (3.35)$$

Using Min terms to write a function from a truth table is simply a matter of combining the Min terms that have a logic High output using an OR gate.

3.2.2.2.   THE PRODUCT OF SUMS FORM

The Product of Sums form gets its name from the way the logic function is written. The POS form uses the Max terms to identify the function. Since each Max term is the result of ORing the complements of the literals, we get the Sum part of the expression. Then all the Max terms are combined using the AND operation, hence the term Products. To write the logic expression in the POS form, we only look at those rows of the truth table that have a logic Low output. We identify these rows as the Max terms and add the Max terms to get the POS form of the logic function. The examples in Figure 3.5 will make this clear.

To write function F_1, we see that we have only four entries that are logic Low output. They correspond to the Max terms M_0, M_1, M_6, and M_7. We will build this function by combining these four rows (Max terms) by an AND gate. So to write the function F_1, we would "AND" the four Max terms that have the logic Low output, as shown in Equation 3.36.

$$F_1 = \prod \left(M_0, M_1, M_6, M_7 \right)$$
$$F_1 = \left(\overline{w} + x + y \right) \bullet \left(w + x + \overline{y} \right) \bullet \left(\overline{w} + \overline{x} + y \right) \bullet \left(\overline{w} + \overline{x} + \overline{y} \right) \quad (3.36)$$

To write function 2, we see that we have only four entries that are logicLow. They correspond to the Max terms M_1, M_2, M_4, and M_7. We will build this function by combining these four rows (Max terms) by an AND gate. So to write the function F_2, we would "Sum" the four Min terms that have the logic Low output, as shown in Equation 3.37.

$$F_2 = \prod \left(M_1, M_2, M_4, M_7 \right)$$
$$F_2 = \left(w + x + \overline{y} \right) \bullet \left(w + \overline{x} + y \right) \bullet \left(\overline{w} + x + y \right) \bullet \left(\overline{w} + \overline{x} + \overline{y} \right) \quad (3.37)$$

Using Max terms to write a function from a truth table is simply a matter of combining the Max terms that have a logic Low output using an AND gate.

A note of interest: Even though Equations 3.34 and 3.36 do not look like each other, they represent the same logic function, so they can be shown to be equivalent to each other. Similarly for Equations 3.35 and 3.37.

3.2.2.3. CONVERTING BETWEEN THE TWO CANONICAL FORMS

We can write any logic function in one of the two canonical forms. The two forms are complementary in the sense that we can easily map the expression for one form into an expression for the other form. Comparing the expressions for the function F_1 in Equations 3.34 and 3.36 gives us the idea.

a. When we are given the Min term expression and want to convert to Max term notation, we first replace the summation symbol with the product symbol. Then we replace the term numbers that are in the Min term notation with those term numbers that are not present using the Max term representation. This is very similar to applying the DeMorgan's Theorem on the Min term expression.

Example: $f(w,x,y) = \sum (m_2, m_3, m_5, m_6) = \prod (M_0, M_1, M_4, M_7)$

b. When we are given the Max term expression and want to convert to Min term notation, we first replace the product symbol with the summation symbol. Then we replace the term numbers that are in the Max term notation with those term numbers that are not present using the Min term representation. This is very similar to applying the DeMorgan's Theorem on the Max term expression.

Example: $f(w,x,y) = \prod (M_0, M_2, M_6, M_7) = \sum (m_1, m_3, m_4, m_5)$

c. When we want to obtain the complement of the function given in the Min term notation, we follow a similar procedure. This time, we have a Min term expression and we want another expression that is also in Min term notation but is the complement of the first. To get the complement function, we replace the terms that are in the Min term expression by the terms that are missing from the Min term expression. The following example shows this.

Example:

$$f(w,x,y) = \sum(m_1, m_3, m_4, m_5)$$

$$\overline{f(w,x,y)} = \sum(m_0, m_2, m_6, m_7)$$

d. When we want to obtain the complement of the function given in the Max term notation, we follow a similar procedure. This time, we have a Max term expression and we want another expression that is also in Max term notation but is the complement of the first. To get the complement function, we replace the terms that are in the Max term expression by the terms that are missing from the Max term expression. The following example shows this.

Example:

$$f(w,x,y) = \prod(M_2, M_3, M_5, M_6)$$

$$\overline{f(w,x,y)} = \prod(M_1, M_2, M_4, M_7)$$

Review Questions for Section 3.2.2

Question: Write the SOP expression for the truth table shown in Figure 3.6 for the two functions F_3 and for function F_4.

Answer: We first write the logic function using the Min terms. Next, we expand the Min terms to get the logic expression. The logic functions are given in Equations 3.38 and 3.39.

$$F_3 = \sum(m_0, m_2, m_4, m_6) = \overline{w} \bullet \overline{x} \bullet \overline{y} + \overline{w} \bullet x \bullet \overline{y} + w \bullet \overline{x} \bullet \overline{y} + w \bullet x \bullet \overline{y}$$

(3.38)

$$F_4 = \sum(m_0, m_4, m_5, m_7) = \overline{w} \bullet \overline{x} \bullet \overline{y} + w \bullet \overline{x} \bullet \overline{y} + w \bullet \overline{x} \bullet y + w \bullet x \bullet y$$

(3.39)

W	X	Y	Function F_3	Function F_4	Min Term	Max Term
0	0	0	1	1	m_0	M_0
0	0	1	0	0	m_1	M_1
0	1	0	1	0	m_2	M_2
0	1	1	0	0	m_3	M_3
1	0	0	1	1	m_4	M_4
1	0	1	0	1	m_5	M_5
1	1	0	1	0	m_6	M_6
1	1	1	0	1	m_7	M_7

FIGURE 3.6. Writing a Logic function from a Truth Table.

Question: Write the POS expression for the truth table shown in Figure 3.6 for the two functions F_3 and for function F_4.

   Answer: We first write the logic function using the Max terms. Next, we expand the Max terms to get the logic expression. The logic functions are given in Equations 3.40 and 3.41.

$$F_3 = \prod \left(M_1, M_3, M_5, M_7 \right) = \left(w + x + \bar{y} \right) \bullet \left(w + \bar{x} + \bar{y} \right) \bullet \left(\bar{w} + x + \bar{y} \right) \bullet \left(\bar{w} + \bar{x} + \bar{y} \right)$$

$$(3.40)$$

$$F_3 = \prod \left(M_1, M_2, M_3, M_6 \right) = \left(w + x + \bar{y} \right) \bullet \left(w + \bar{x} + y \right) \bullet \left(w + \bar{x} + \bar{y} \right) \bullet \left(\bar{w} + \bar{x} + y \right)$$

$$(3.41)$$

Question: Using the SOP expression for the truth table shown in Figure 3.6 for the two functions F_3 and for function F_4, write the complement of the function in the SOP format.

   Answer: Following the guideline given, we can write the complements in the SOP format as shown in Equation 3.42.

$$F_3 = \sum \left(m_0, m_2, m_4, m_6 \right) \Rightarrow \overline{F_3} = \sum \left(m_1, m_3, m_5, m_7 \right)$$

$$F_4 = \sum \left(m_0, m_4, m_5, m_7 \right) \Rightarrow \overline{F_4} = \sum \left(m_1, m_2, m_3, m_6 \right)$$

$$(3.42)$$

3.3. BUILDING LOGIC FUNCTIONS FROM TRUTH TABLES

Sometimes we do not have a logic equation to represent a logic function but we have the truth table of the logic function. In these cases, we have to first write a logic equation before we can minimize the logic equation to get a minimum expression. The procedure to do this is quite simple and straightforward.

3.3.1. SUM OF PRODUCTS EXPRESSIONS (SOP)

To write the logic function in the SOP format, we look for the rows of the truth table that are logic High. For each logic High output, we determine the logic expression by combining the logic variables using the AND operation. If the logic variable in the truth table is High, then we write the variable as it is. If the variable in the truth table is Low, then we write the variable as a complement. Once we have the expressions for each logic High output, we combine all the expressions using the OR operation. Each one of the rows of the truth table represents a Min term of the logic function. The logic function that we form in this way gives us the Min term canonical form. Examine the two truth tables in Figure 3.7. The first truth table is for a completely specified function, and its logic function is given in Figure 3.7.

The second truth table is for an incompletely specified function, so this time we have some don't care terms present. We form the expressions for the don't care terms in the same way, but since these terms are don't care, we identify this by placing all these terms inside brackets with a big "D" before the logic expressions. From this logic function, we can write the canonical expression using the Min terms. This is also shown in Figure 3.7.

W	X	Y	F
0	0	0	0
0	0	1	0
0	1	0	1
0	1	1	0
1	0	0	1
1	0	1	0
1	1	0	1
1	1	1	1

W	X	Y	G
0	0	0	1
0	0	1	x
0	1	0	1
0	1	1	0
1	0	0	0
1	0	1	x
1	1	0	x
1	1	1	1

$$F = \bar{W} \bullet \bar{X} \bullet \bar{Y} + W \bullet \bar{X} \bullet \bar{Y} + W \bullet X \bullet \bar{Y} + W \bullet X \bullet Y = \sum (m_2, m_4, m_6, m_7)$$

$$G = \bar{W} \bullet \bar{X} \bullet \bar{Y} + \bar{W} \bullet X \bullet \bar{Y} + W \bullet X \bullet Y + D(\bar{W} \bullet \bar{X} \bullet Y + W \bullet \bar{X} \bullet Y + W \bullet X \bullet \bar{Y})$$

$$= \sum (m_0, m_2, m_7) + (d_1, d_5, d_6)$$

FIGURE 3.7. Writing Logic expressions from Truth Tables.

3.3.2. PRODUCT OF SUMS EXPRESSIONS (POS)

To write the logic function in the POS format, we look for the rows of the truth table that are logic Low. For each logic Low output, we determine the logic expression by combining the logic variables using the OR operation. If the logic variable in the truth table is High, then we write the complement of that variable. If the variable in the truth table is Low, then we write the variable as it is. Once we have the expressions for each logic Low output, we combine all the expressions using the AND operation. Each one of the rows of the truth table represents a Max term of the logic function. The logic function that we form in this way gives us the Max term canonical form. Examine the two truth tables in Figure 3.8. The first truth table is for a completely specified function, and its logic function is given in Figure 3.8. The second truth table is for an incompletely specified function, so this time we have some don't care terms present. We form the expressions for

W	X	Y	F		W	X	Y	G
0	0	0	0		0	0	0	1
0	0	1	0		0	0	1	x
0	1	0	1		0	1	0	1
0	1	1	0		0	1	1	0
1	0	0	1		1	0	0	0
1	0	1	0		1	0	1	x
1	1	0	1		1	1	0	x
1	1	1	1		1	1	1	1

$$F = \left(W + X + Y\right) \cdot \left(W + X + \overline{Y}\right) \cdot \left(W + \overline{X} + \overline{Y}\right) \cdot \left(\overline{W} + X + \overline{Y}\right) = \prod M_0, M_1, M_3, M_5$$

$$G = \left(W + \overline{X} + \overline{Y}\right) \cdot \left(\overline{W} + X + Y\right) + \mathcal{D}\left(\left(W + X + \overline{Y}\right) \cdot \left(\overline{W} + X + \overline{Y}\right) \cdot \left(\overline{W} + \overline{X} + Y\right)\right)$$

$$= \sum \left(M_3, M_4\right) + \left(D_1, D_5, D_6\right)$$

FIGURE 3.8. Writing Logic expressions from Truth Tables.

the don't care terms in the same way, but since these terms are don't care, we identify this by placing all these terms inside brackets with a big "D" before the logic expressions. From this logic function, we can write the canonical expression using the Min terms. This is also shown in Figure 3.8.

3.4.   CHAPTER PROBLEMS

3.4.1.   Verify the Distributive Law for Boolean Algebra.
3.4.2.   Verify the Commutative Law for Boolean Algebra.
3.4.3.   Verify the Associative Law for Boolean Algebra.
3.4.4.   Prove the following using the laws and theorems of Boolean Algebra.
  3.4.4.1. $\overline{A} \cdot \overline{B} + \overline{A} \cdot B = \overline{A}$ and $\left(\overline{A} + \overline{B}\right) \cdot \left(\overline{A} + B\right) = \overline{A}$

3.4.4.2. $\bar{A} + \bar{A} \bullet B = \bar{A}$ and $\left(\bar{A}\right) \bullet \left(\bar{A} + B\right) = \bar{A}$

3.4.4.3. $\left(\bar{B} + \bar{A}\right) \bullet B = \bar{A} \bullet B$ and $\left(\bar{A} \bullet \bar{B}\right) + \left(B\right) = \bar{A} + B$

3.4.5.   Use the laws of Boolean Algebra to obtain a lower-cost SOP expression.

3.4.5.1.
$$F = \bar{A} \bullet \bar{B} \bullet \bar{C} \bullet \bar{D} + \bar{A} \bullet B \bullet \bar{C} \bullet \bar{D} + A \bullet B \bullet C \bullet \bar{D}$$
$$+ A \bullet \bar{B} \bullet C \bullet D + \bar{A} \bullet \bar{B} \bullet C \bullet D + A \bullet B \bullet C \bullet D$$

3.4.5.2.
$$G = \bar{A} \bullet B \bullet \bar{C} \bullet \bar{D} + \bar{A} \bullet B \bullet \bar{C} \bullet D + A \bullet \bar{B} \bullet \bar{C} \bullet D$$
$$+ A \bullet \bar{B} \bullet C \bullet D + \bar{A} \bullet \bar{B} \bullet C \bullet \bar{D} + \bar{A} \bullet B \bullet C \bullet \bar{D}$$

3.4.5.3.
$$F = \bar{A} \bullet \bar{B} \bullet \bar{C} \bullet D + \bar{A} \bullet B \bullet \bar{C} \bullet D + A \bullet B \bullet \bar{C} \bullet D$$
$$+ A \bullet \bar{B} \bullet C \bullet D + A \bullet \bar{B} \bullet C \bullet \bar{D} + A \bullet B \bullet C \bullet D$$

3.4.5.4.
$$G = \bar{A} \bullet \bar{B} \bullet \bar{C} \bullet \bar{D} + A \bullet B \bullet \bar{C} \bullet \bar{D} + A \bullet \bar{B} \bullet \bar{C} \bullet \bar{D}$$
$$+ \bar{A} \bullet \bar{B} \bullet C \bullet D + \bar{A} \bullet \bar{B} \bullet C \bullet \bar{D} + A \bullet B \bullet C \bullet \bar{D}$$

3.4.6.   Determine an expression for the complement of the following functions:

3.4.6.1. $f_{(A,B,C,D)} = \left[A + \left(\overline{BCD}\right)\right]\left[\overline{AD} + B\left(\bar{C} + A\right)\right]$

3.4.6.2. $f_{(A,B,C,D)} = AB\bar{C} + \left(A + \bar{B} + D\right)\left(\bar{A}BD + \bar{B}\right)$

3.4.7.   Using laws of Boolean Algebra, determine which logic gate the expression in Equation 3.43 represents.

$$\left(X \bullet \left(X \bullet Y\right)\right) \bullet \left(Y \bullet \left(X \bullet Y\right)\right) \tag{3.43}$$

4. SIMPLIFYING A LOGIC FUNCTION

4.0.   INTRODUCTION

Earlier, we saw that we can simplify a logic function using the laws and the theorems of Boolean Algebra. I am sure you realized that using the laws of Boolean Algebra did not show us any algorithm that you could use to simplify the function. You did not know when you had the minimum expression. Another difficulty we saw was that we had to make the function more complex before we got the simplification that we wanted. This happened when we introduced the expression $(x + \overline{x}) = 1$ in the logic function to introduce a variable that was missing. Finally, it is too cumbersome and error-prone to use the Boolean Algebra to simplify a logic function.

Against the above scenario, we can look at computer-based algorithm methods that will simplify a logic function. The computer-based algorithms not only work very nicely with functions of a few variables (like three or four), but they also extend to functions of many more variables very nicely. In fact, there is no limit to the number of variables that a computer-based algorithm can handle with relative ease.

In what follows, we will first look at a graphical technique to minimize a logic function. This method is used by humans who have to simplify a relatively small logic function quickly. Next, we will look at an algorithm method that can be used on a computer to simplify a logic function. In this chapter, we also introduce the concept of don't cares in a logic function.

4.1. THE K-MAP METHOD

The method that we are going to examine in this section is known as the "Karnaugh map" method. This method is a graphical method. To understand the method, we first need to learn how we will be drawing the graph and what the graph represents. To do this, we begin with a map of one variable, which is drawn in Figure 4.1.

In the one-variable K-map, we see two squares. There are two squares because a single variable can take on only two unique values, so on a one-variable K-map, each square represents the value that the variable can take on. The way the map is drawn represents a function of variable "x." The square with a "0" written in it represents the value of the function when the variable x is zero, and the square with a "1" written in it represents the value

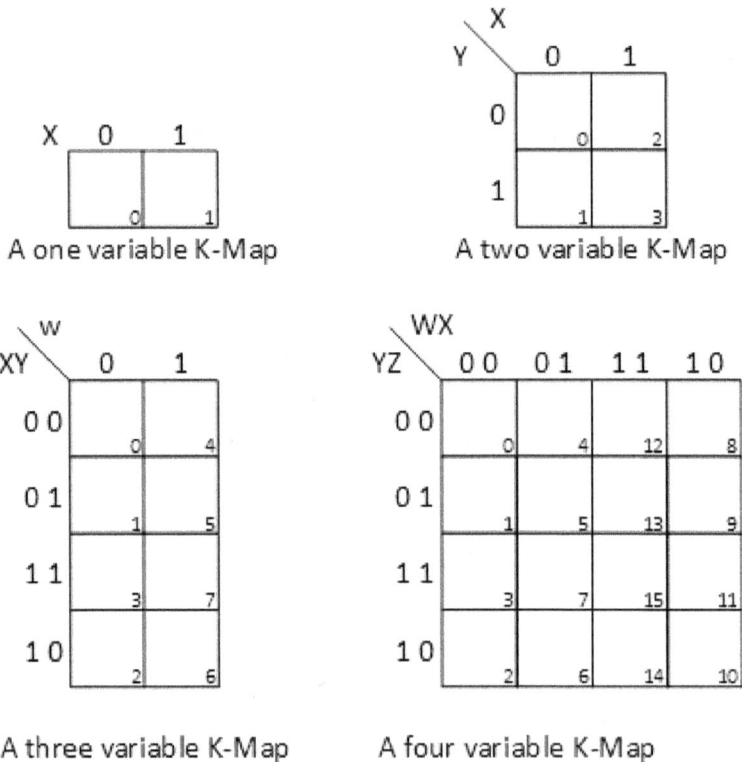

A one variable K-Map

A two variable K-Map

A three variable K-Map

A four variable K-Map

FIGURE 4.1. A one, two, three, and a four variable K-Map.

of the function when the variable x is one. That is all there is to this method. Each square represents some combinations of the variables that make up the function, and the value inside the square represents the output from the logic function when the variables take on that value.

Next, we look at the two-variable K-map. This time, since we have two variables, we have shown the map in a compact manner with one variable across the top and the other one down the side. In this map, the variable X is zero in the squares with "0" and "1" written in them. The variable X is zero for this entire column under the value 0 for the variable X. The variable X is one for the other entire column. Similarly, the variable Y is zero in the squares with "0" and "2" written in them. The variable Y is zero for this entire row that is headed by the value for the variable Y as 0 and the variable Y is one for the other entire row. When we combine both the variables for any one square, we get the logic function that the square represents. For example, the square with a 1 written in it represents the logic function $\overline{x} \bullet y$ since the value of the variable x is zero in this square and the value of the variable y is 1 in this square.

We can extend this idea to three- and four-variable maps also. In both the three- and the four-variable maps, we have labeled the rows and the columns in a special manner. The labeling is not straight binary. We have labeled the rows and columns in "Gray code." Gray code is discussed in Appendix 4.a of this chapter. Just as in the K-maps of one and two variables, each of the squares of the K-maps of three and four variables also represents a logic function.

In the three-variable map, the square with "3" written in it represents the logic function $\overline{w} \bullet x \bullet y$, and in the four-variable map, the square with a "9" written in it represents the logic function $w \bullet \overline{x} \bullet \overline{y} \bullet z$.

We have established that each square in the K-map represents a unique combination of the logic variables. The relation that the truth table has with a logic function is the same relation that the K-map has with the logic function. Each row of the truth table represents a unique combination of the input variables and tells us what to expect as output from the logic function when the variables have those values. Each square of the K-map represents a unique combination of the input variables and tells us what to expect as output from the logic function by what is written in the square of the K-map when the variables have those values.

4.1.1. USING ADJACENT SQUARES
ON THE K-MAP

The simplest of all theorems of Boolean Algebra is the complements theorem. It tells us how the complements combine. This theorem is given in Equation 3.13 and repeated here in Equation 4.1.

$$X + \bar{X} = 1 \qquad\qquad X \bullet \bar{X} = 0 \qquad\qquad (4.1)$$

The Complement Theorem helps us to eliminate one of the variables from a logic expression when the condition is right. Another theorem we often use is the Simplification Theorem a. The two theorems together form the basis of how simplification works when we use the K-map. The Simplification Theorem is given in Equation 3.16a and repeated here in Equation 4.2.

$$X \bullet Y + X \bullet \bar{Y} = X \qquad\qquad \left(X + Y\right) \bullet \left(X + \bar{Y}\right) = X \qquad (4.2)$$

The Simplification Theorem allows us to isolate a variable and its complement and then the Complement Theorem allows us to eliminate that variable. We will now apply these ideas to the function given in Figure 4.2.

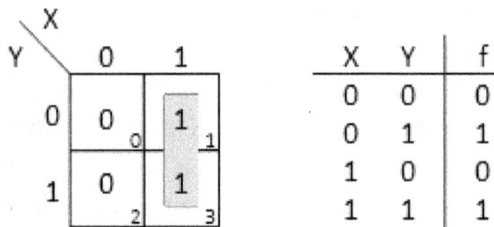

FIGURE 4.2. Using a K-map to minimize a function.

From the truth table, we can minimize the function as shown in Equation 4.3.

$$f = x \bullet \bar{y} + x \bullet y$$
$$f = x \bullet (\bar{y} + y) \qquad \text{Simplification Theorem a}$$
$$f = x \bullet (1) = x \qquad \text{Complements Theorem}$$

$$(4.3)$$

We use the exact same process when we use the K-map. In the K-map, any two squares that are adjacent to each other (share an edge) have this property. Looking at the K-map in Figure 4.2, we see that the square identified with the number "1" represents the logic function $x \bullet \bar{y}$ and the square identified with the number "3" represents the logic function $x \bullet y$. We also see that the two squares in the K-map share an edge and hence are adjacent to each other. To $\bar{X}$ write the logic expression that is represented by the two adjacent squares, we group them together by drawing a ring around the two entries in the K-map. Now that we have identified the two squares to be combined together, we write out only those variables that are the same in the two adjacent squares, as shown in Figure 4.2. This leads us to the required expression, just as we did in Equation 4.3.

As another example, determine the logic expression represented by the truth table in Figure 4.3. We first transfer the logic values from the truth table to the K-map. This is done in Figure 4.3. Next, we look for adjacent

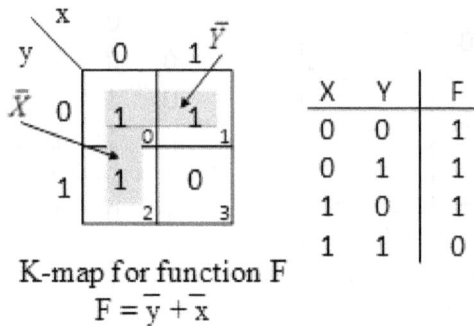

K-map for function F
$$F = \bar{y} + \bar{x}$$

X	Y	F
0	0	1
0	1	1
1	0	1
1	1	0

FIGURE 4.3. Using a K-map to minimize a function.

squares with 1 in them. We see that squares "0" and "1" share an edge and hence are adjacent to each other; squares "0" and "2" also share an edge, so they are also adjacent to each other. All three squares have a 1 entered in the square. Writing the logic expression for the function, we get Equation 4.4.

$$f = \overline{x} \bullet \overline{y} + x \bullet \overline{y} + \overline{x} \bullet y$$
$$f = (\overline{x} + x) \bullet \overline{y} + \overline{x} \bullet (\overline{y} + y) \qquad (4.4)$$
$$f = \overline{y} + \overline{x}$$

4.1.2. ADJACENCIES OF HIGHER DIMENSIONS

Now that we have seen how to recognize two squares in a K-map that are adjacent to each other, let us extend this idea to recognize adjacencies of higher dimensions. Examine Figure 4.4. It represents a function of three variables. In Figure 4.4, we have identified two groups of two Min terms each. The two Min terms in each of the two groups are adjacent to each other; hence, the first group covering squares 1 and 3 can be written as $Y \bullet \overline{Z}$ and the second group covering squares 5 and 7 can be written as $Y \bullet Z$. When we examine the two expressions for the two groups, we immediately see that we can use

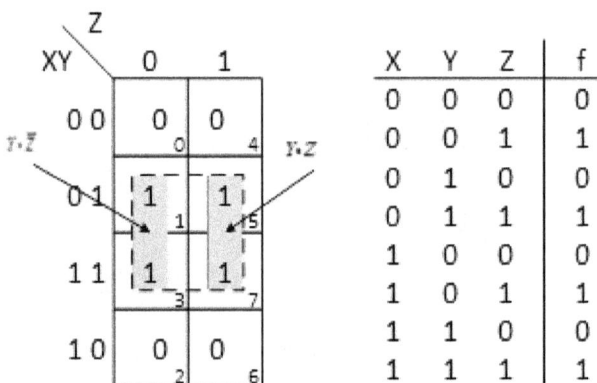

X	Y	Z	f
0	0	0	0
0	0	1	1
0	1	0	0
0	1	1	1
1	0	0	0
1	0	1	1
1	1	0	0
1	1	1	1

FIGURE 4.4. Using a K-map to minimize a function.

the Simplification Theorem a on the two groups and then the Complements Theorem on the two groups, as shown in Equation 4.5.

$$Y \bullet \overline{Z} + Y \bullet Z = Y \bullet \left(\overline{Z} + Z \right) = Y \qquad (4.5)$$

Equation 4.5 shows us how we can combine four terms together to form one group. The result of applying the Simplification and the Complements theorems on the groups is the elimination of another variable. To see how we can recognize this group of four terms that can be combined in a group, we examine the characteristics of the function in Figure 4.4 as follows.

a. Find two squares that are adjacent to each other. Combine into one group. The two squares that form a group have one side that is adjacent to each other.

b. Find two other squares that are adjacent to each other. Combine into a different group. These two items that form this second group have one side that is adjacent to each other.

c. If the two groups that we just formed also share a side (but not any squares), then the two groups are also adjacent to each other and hence can be combined together to form a bigger group. The side shared is the side of the group, so the entire side of one group must be shared by the entire side of the other group.

The characteristics of this larger group are the same as the characteristics of the smaller group. These are: All the squares that combine in forming this larger group have one or more variables that are identical to all the members of the group. In Figure 4.4, this is variable Y. All the other variables, the variables that are eliminated, appear in their **true** form half the time and in their **complemented** form half the time. In Figure 4.4, this is the variable X and the variable Z. Finally, to write the expression for the group we AND together the variables that are the same in all the squares of the group.

Question: In Figure 4.1, what is the logic function represented by the square identified with the number 1 in the square?

    Answer: For the K-map of the one-variable function, the square represents the function X.

    For the K-map of the two-variable function, the square represents the function $\bar{X} \bullet Y$.

    For the K-map of the three-variable function, the square represents the function $\bar{W} \bullet \bar{X} \bullet Y$.

    For the K-map of the four-variable function, the square represents the function $\bar{W} \bullet \bar{X} \bullet \bar{Y} \bullet Z$.

Question: In a four-variable K-map, identify the squares that would represent the following Min terms: m_3, m_7, m_8, m_{11}, m_{13}, m_{15}.

    Answer: The squares in a four-variable K-map that would represent the required Min terms are the squares with 3, 7, 8, 11, 13, and 15 written in them.

4.2.   USING TWO-, THREE-, AND FOUR-VARIABLE K-MAPS

Now that we know how to recognize adjacent squares and combine them together to form a minimum logic function, we will apply this method to see how it works. As we study this method, you will find that the K-maps are much easier to use than the Boolean Algebra method and we get the result that we would expect. In addition to being efficient, this method has a second advantage that we will see toward the end of this section. This advantage is the detection and the elimination of hazards.

K-map for function F_1
$F_1 = \bar{X}$

K-map for function F_2
$F_2 = \bar{Y}$

K-map for function F_3
$F_3 = \bar{X} \cdot Y + X \cdot \bar{Y}$

X	Y	F_1	F_2	F_3
0	0	1	1	0
0	1	1	0	1
1	0	0	1	1
1	1	0	0	0

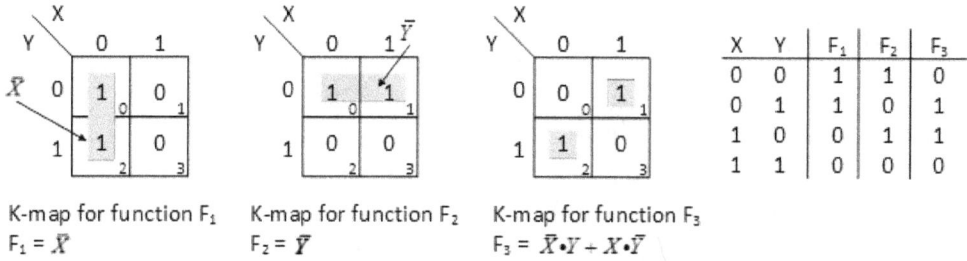

FIGURE 4.5. Using a K-map to minimize a function.

4.2.1. MINIMUM SUM OF PRODUCTS EXPRESSIONS

Examine the truth table and the K-maps for functions F_1, F_2, and F_3 in Figure 4.5. These are all two-variable K-maps. For function F_1, look at Equation 4.6.

$$\bar{X} \bullet \bar{Y} + \bar{X} \bullet Y = \bar{X} \bullet \left(\bar{Y} + Y\right) = \bar{X} \qquad (4.6)$$

In example function F_1, we have two adjacent squares. These two adjacent squares can be combined as shown in Equation 4.6. Next, for function F_2, look at Equation 4.7.

$$\bar{X} \bullet \bar{Y} + X \bullet \bar{Y} = \bar{Y} \bullet \left(\bar{X} + X\right) = \bar{Y} \qquad (4.7)$$

In example function F_2, we have two adjacent squares that have logic High entered. The two squares can be combined as shown in Equation 4.7. Next, for function F_3, look at Equation 4.8.

$$\bar{X} \bullet Y + X \bullet \bar{Y} \qquad (4.8)$$

In example function F_3, we do not have any adjacent squares as no squares share a common edge. Since no squares are adjacent, the function can be written as shown in Equation 4.8.

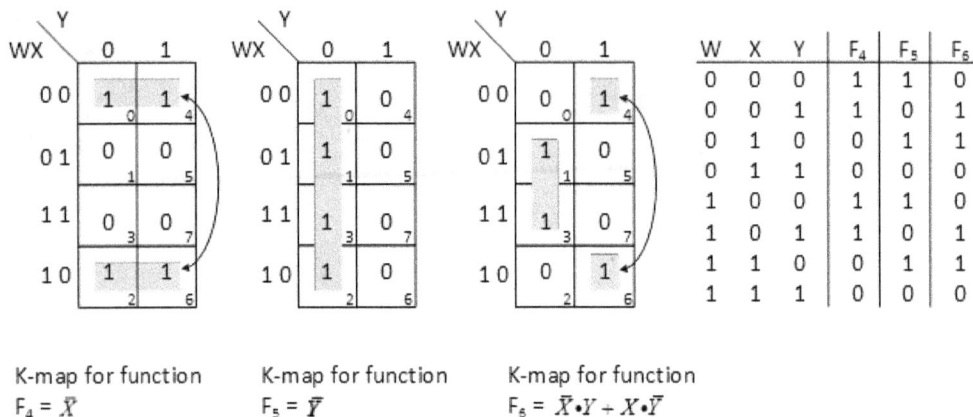

K-map for function $F_4 = \overline{X}$

WX \ Y	0	1
0 0	1	1
0 1	0	0
1 1	0	0
1 0	1	1

K-map for function $F_5 = \overline{Y}$

WX \ Y	0	1
0 0	1	0
0 1	1	0
1 1	1	0
1 0	1	0

K-map for function $F_6 = \overline{X} \bullet Y + X \bullet \overline{Y}$

WX \ Y	0	1
0 0	0	1
0 1	1	0
1 1	1	0
1 0	0	1

W	X	Y	F_4	F_5	F_6
0	0	0	1	1	0
0	0	1	1	0	1
0	1	0	0	1	1
0	1	1	0	0	0
1	0	0	1	1	0
1	0	1	1	0	1
1	1	0	0	1	1
1	1	1	0	0	0

FIGURE 4.6. Using a K-map to minimize a function.

Examine the truth table and the K-maps for functions F_4, F_5, and F_6 in Figure 4.6. These are all three-variable K-maps. For function F_4, look at Equation 4.9.

$$F_4 = \left(\overline{W} \bullet \overline{X} \bullet \overline{Y} + \overline{W} \bullet \overline{X} \bullet Y\right) + \left(W \bullet \overline{X} \bullet \overline{Y} + W \bullet \overline{X} \bullet Y\right)$$
$$F_4 = \left(\overline{W} \bullet \overline{X} \bullet \left(\overline{Y} + Y\right)\right) + \left(W \bullet \overline{X} \bullet \left(\overline{Y} + Y\right)\right) \quad (4.9)$$
$$F_4 = \left(\overline{W} \bullet \overline{X}\right) + \left(W \bullet \overline{X}\right) = \left(\overline{W} + W\right) \bullet \overline{X} = \overline{X}$$

In example function F_4, we have two groups of two adjacent squares. The two squares of each of these groups can be combined as shown in Equation 4.9. After we combine the squares together to form the two groups, we see that the two groups now share a common edge. Since the groups share a common edge, we can combine the two groups together into a larger group. This is also shown in Equation 4.9. For function F_5, look at Equation 4.10.

$$F_5 = \left(\overline{W} \bullet \overline{X} \bullet \overline{Y} + \overline{W} \bullet X \bullet \overline{Y}\right) + \left(W \bullet X \bullet \overline{Y} + W \bullet \overline{X} \bullet \overline{Y}\right)$$
$$F_5 = \left(\overline{W} \bullet \overline{Y} \bullet \left(\overline{X} + X\right)\right) + \left(W \bullet \overline{Y} \bullet \left(X + \overline{X}\right)\right) \quad (4.10)$$
$$F_5 = \left(\overline{W} \bullet \overline{Y}\right) + \left(W \bullet \overline{Y}\right) = \left(\overline{W} + W\right) \bullet \overline{Y} = \overline{Y}$$

In example function F_5, we have two groups of two adjacent squares. The two squares of each of these groups can be combined as shown in Equation 4.10. After we combine the squares together to form the two groups, we see that the two groups share a common edge. Since the groups share a common edge, we can combine the two groups together into a larger group. This is also shown in Equation 4.10. For function F_6, look at Equation 4.11.

$$F_6 = \left(\overline{W} \bullet X \bullet \overline{Y} + W \bullet X \bullet \overline{Y} \right) + \left(\overline{W} \bullet \overline{X} \bullet Y + W \bullet \overline{X} \bullet Y \right)$$
$$F_6 = \left(X \bullet \overline{Y} \bullet \left(\overline{W} + W \right) \right) + \left(\overline{X} \bullet Y \bullet \left(\overline{W} + W \right) \right) \qquad (4.11)$$
$$F_6 = \left(X \bullet \overline{Y} \right) + \left(\overline{X} \bullet Y \right)$$

In example function F_6, we have two groups of two adjacent squares. The two squares of each of these groups can be combined as shown in Equation 4.10. Since no more squares or groups of squares are adjacent to each other, the function can be written as shown in Equation 4.11.

Examine the truth table for functions F_7 and F_8 in Figure 4.7. Both of these are four-variable K-maps. For function F_7, look at Equation 4.12.

$$F_7 = \overline{W} \bullet \overline{X} \bullet \overline{Y} \bullet \overline{Z} + \overline{W} \bullet \overline{X} \bullet Y \bullet Z + \overline{W} \bullet X \bullet \overline{Y} \bullet \overline{Z} + \overline{W} \bullet X \bullet \overline{Y} \bullet Z$$
$$+ W \bullet \overline{X} \bullet \overline{Y} \bullet \overline{Z} + W \bullet X \bullet \overline{Y} \bullet Z + W \bullet X \bullet Y \bullet Z + W \bullet \overline{X} \bullet Y \bullet \overline{Z}$$
$$F_7 = \overline{W} \bullet \overline{Y} \bullet \overline{Z} \bullet \left(\overline{X} + X \right) + \overline{W} \bullet X \bullet \overline{Y} \bullet \left(\overline{Z} + Z \right) + W \bullet X \bullet Z \bullet \left(\overline{Y} + Y \right) \quad (4.12)$$
$$+ W \bullet \overline{X} \bullet \overline{Z} \bullet \left(\overline{Y} + Y \right) + \overline{W} \bullet \overline{X} \bullet Y \bullet Z$$
$$F_7 = \overline{W} \bullet \overline{Y} \bullet \overline{Z} + \overline{W} \bullet X \bullet \overline{Y} + W \bullet X \bullet Z + W \bullet \overline{X} \bullet \overline{Z} + \overline{W} \bullet \overline{X} \bullet Y \bullet Z$$

In example function F_7, we have four groups of two adjacent squares and one more square that is not adjacent to any other square that has a 1 entered in it. None of these groups can be combined together as they do not share an edge. Note that the groups formed by the squares 0,4 and squares 4,5 share a square, but they do not share an edge, so they cannot be combined together into a larger group. The result for function F_7 is shown in Equation 4.12. For function F_8, look at Equation 4.13.

WX

YZ	00	01	11	10
00	1 (0)	1 (4)	(12)	1 (8)
01	(1)	1 (5)	1 (3)	(9)
11	1 (3)	(7)	1 (15)	(11)
10	(2)	(6)	(14)	1 (10)

K-map for function

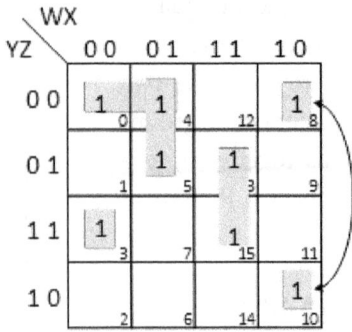

$$F_7 = \bar{Z}\cdot\bar{Y}\cdot\bar{W} + \bar{Y}\cdot\bar{W}\cdot X + \bar{Z}\cdot W\cdot\bar{X} + Z\cdot Y\cdot\bar{W}\cdot\bar{X}$$

W	X	Y	Z	F_7	F_8
0	0	0	0	1	1
0	0	0	1	0	0
0	0	1	0	0	1
0	0	1	1	1	0
0	1	0	0	1	1
0	1	0	1	1	0
0	1	1	0	0	1
0	1	1	1	0	0
1	0	0	0	1	0
1	0	0	1	0	0
1	0	1	0	1	1
1	0	1	1	0	1
1	1	0	0	0	0
1	1	0	1	1	0
1	1	1	0	0	1
1	1	1	1	1	0

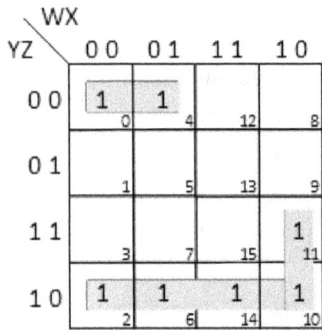

WX

YZ	00	01	11	10
00	1 (0)	1 (4)	(12)	(8)
01	(1)	(5)	(13)	(9)
11	(3)	(7)	(15)	1 (11)
10	1 (2)	1 (6)	1 (14)	1 (10)

Use the K-map to minimize the given functions

K-map for function

$$F_8 = \bar{W}\cdot\bar{Z} + W\cdot\bar{X}\cdot Y + Y\cdot\bar{Z}$$

FIGURE 4.7. Using a K-map to minimize a function.

$$
\begin{aligned}
F_8 &= \bar{W}\bullet\bar{X}\bullet\bar{Y}\bullet\bar{Z} + \bar{W}\bullet X\bullet\bar{Y}\bullet\bar{Z} + \bar{W}\bullet\bar{X}\bullet Y\bullet\bar{Z} + \bar{W}\bullet X\bullet Y\bullet\bar{Z} \\
&\quad + W\bullet X\bullet Y\bullet\bar{Z} + W\bullet\bar{X}\bullet Y\bullet\bar{Z} + W\bullet\bar{X}\bullet Y\bullet Z \\
F_8 &= \bar{W}\bullet\bar{Y}\bullet\bar{Z}\bullet\left(\bar{X}+X\right) + \bar{W}\bullet Y\bullet\bar{Z}\bullet\left(\bar{X}+X\right) \\
&\quad + W\bullet Y\bullet\bar{Z}\bullet\left(\bar{X}+X\right) + W\bullet\bar{X}\bullet Y\bullet\left(\bar{Z}+Z\right) \\
F_8 &= \bar{W}\bullet\bar{Z}\bullet\left(\bar{Y}+Y\right) + Y\bullet\bar{Z}\bullet\left(\bar{W}+W\right) + W\bullet\bar{X}\bullet Y \\
F_8 &= \bar{W}\bullet\bar{Z} + Y\bullet\bar{Z} + W\bullet\bar{X}\bullet Y
\end{aligned}
\tag{4.13}
$$

In example function F_8, we have four groups of two adjacent squares. Two of these groups, consisting of squares 2 and 6 and squares 10 and 14, also share an edge. The four squares that make up these two groups can be combined into one larger group. These two groups are combined into a group of four terms. Two other groups, consisting of squares 0 and 4 and squares 2 and 6, also share an edge. These four squares can also be combined into a larger group of four terms. Doing this gives us the final expression in Equation 4.13.

4.2.2. MINIMUM PRODUCT OF SUMS EXPRESSIONS

Until now, we have always written the expression in the SOP form. The K-map can also be used to write the lowest cost expression in the POS form. The procedure is very similar to the procedure that we used for the SOP expression. This time, however, we will look for the 0's in the expression instead of the 1's in the expression. To form a POS expression, we will group together squares that have 0's entered in them. There are a couple of differences in the final expression. The differences are listed below.

a. We interpret the variables as complements, so if a group or a term has a variable in its true form, we write it in the POS expression in its complemented form. If the variable is in its complement form, then we write it in its true form.

b. Each term is formed by ORing the individual variables together.

c. The entire function is formed by ANDing all the individual terms together.

To see how this method works, let us consider functions F_4, F_5, and F_6 given in Figure 4.6 again; this time, we will write the POS expression for the three functions. The three functions are shown in Equation 4.14.

$$F_4 = \left(Y + W + \bar{X}\right) \bullet \left(\bar{Y} + W + \bar{X}\right) \bullet \left(Y + \bar{W} + \bar{X}\right) \bullet \left(\bar{Y} + \bar{W} + \bar{X}\right)$$
$$F_4 = \left(Y \bullet \bar{Y} + W + \bar{X}\right)\left(Y \bullet \bar{Y} + \bar{W} + \bar{X}\right)$$
$$F_4 = \left(W \bullet \bar{W} + \bar{X}\right) = \bar{X}$$

$$F_5 = \left(\bar{Y} + W + X\right) \bullet \left(\bar{Y} + W + \bar{X}\right) \bullet \left(\bar{Y} + \bar{W} + \bar{X}\right) \bullet \left(\bar{Y} + \bar{W} + X\right) \qquad (4.14)$$
$$F_5 = \left(\bar{Y} + W + X \bullet \bar{X}\right)\left(\bar{Y} + \bar{W} + \bar{X} \bullet X\right)$$
$$F_5 = \left(\bar{Y} + W \bullet \bar{W}\right) = \bar{Y}$$

$$F_6 = \left(Y + W + X\right) \bullet \left(\bar{Y} + W + \bar{X}\right) \bullet \left(\bar{Y} + \bar{W} + \bar{X}\right) \bullet \left(Y + \bar{W} + X\right)$$
$$F_6 = \left(Y + W \bullet \bar{W} + X\right) \bullet \left(\bar{Y} + W \bullet \bar{W} + \bar{X}\right)$$
$$F_6 = \left(Y + X\right) \bullet \left(\bar{Y} + \bar{X}\right)$$

Another way to verify this is to write the SOP for the 0's in the function and then apply the DeMorgan's Theorem on the result. Verify this for functions F_4, F_5, and F_6.

Review Questions for Section 4.2

Question: Write the SOP and the POS expressions for the two three-variable functions and the two four-variable functions shown in Figure 4.8.

Answer: The SOP expression for the K-map shown in Figure 4.8a is $\bar{W} \bullet \bar{Y} + W \bullet Y$.

The SOP expression for the K-map shown in Figure 4.8b is $X \bullet \bar{Y} + W \bullet X$.

The SOP expression for the K-map shown in Figure 4.8c is $\bar{X} \bullet \bar{Z} + X \bullet \bar{Y} \bullet Z$.

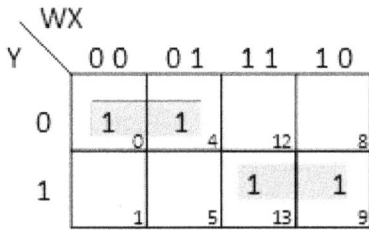

Figure 4.8a

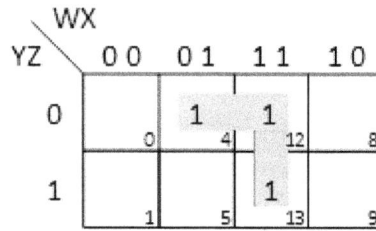

Figure 4.8b

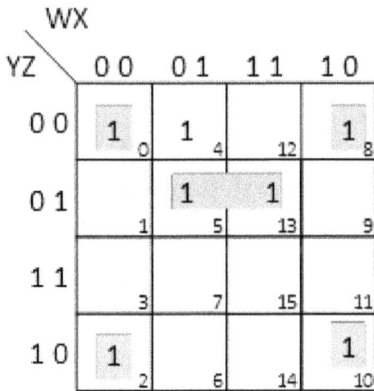

Figure 4.8c

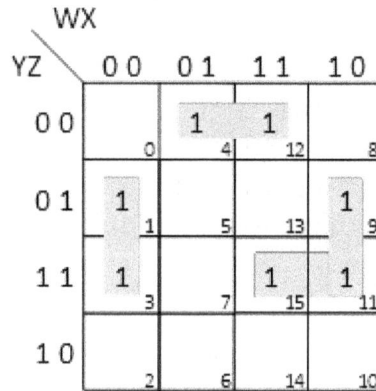

Figure 4.8d

FIGURE 4.8. Using a K-map to minimize a functions.

The SOP expression for the K-map shown in Figure 4.8d is $\bar{X} \bullet Z + X \bullet \bar{Y} \bullet \bar{Z} + W \bullet Y \bullet Z$.

Question: In a four-variable K-map, identify the squares that would represent the following Min terms: m_3, m_7, m_8, m_{11}, m_{13}, m_{15}.

Answer: The squares of a four-variable K-map that would represent the required Min terms are the squares with 3, 7, 8, 11, 13, and 15 written in them.

4.3. DON'T CARE TERMS IN K-MAPS

Until now, we have represented each row in the truth table as either logic High or as logic Low. Sometimes there are occasions when we do not care what the output of the logic function is when the input represents that particular row of the truth table. This often happens when that particular combination of the variables will never appear in the logic function. In these cases, we say that this particular row of the truth table is a don't care. Don't care terms are represented in the canonical form, as shown in Equation 4.15. We can have don't care terms in either the SOP expression or the POS expression. We generally use the Min term or the Max term representation to identify the don't care terms. In the Min term notation, a don't care is indicated as d_0 to represent the Min term m_0 as a don't care term. Similarly, don't care terms in Max term notation are indicated as D_0 to represent the Max term M_0 as a don't care term.

$$\text{SOP expression } f_{WXY} = \sum(m_3, m_4, m_6, m_7) + (d_0, d_5)$$
$$\text{POS expression } f_{WXY} = \prod(M_3, M_4, M_6, M_7) \bullet (D_0, D_5)$$

$$(4.15)$$

To see this in operation, let us look at the truth table in Figure 4.9. This truth table has four inputs and two outputs. It is an *incompletely specified* function because it contains don't care terms.

We first see how we will write the two functions in their canonical representations. Equation 4.16 shows how we write the function F_9, and Equation 4.17 shows us how we write the function F_{10}. This time, we have written each of the functions in its SOP and its POS canonical form.

$$F_9 = \sum(m_0, m_3, m_4, m_8, m_{10}, m_{13}, m_{15}) + (d_1, d_5, d_9)$$
$$F_9 = \prod(M_2, M_6, M_7, M_{11}, M_{12}, M_{14}) \bullet (D_1, D_5, D_9)$$

$$(4.16)$$

$$F_{10} = \sum(m_0, m_2, m_6, m_{10}, m_{14}) + (d_4, d_{11}, d_{12})$$
$$F_{10} = \prod(M_1, M_3, M_5, M_7, M_8, M_9, M_{13}, M_{15}) \bullet (D_4, D_{11}, D_{12})$$

$$(4.17)$$

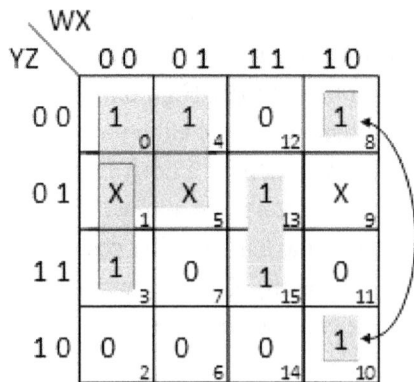

K-map for function

$$F_9 = \overline{W} \cdot \overline{Y} + \overline{W} \cdot \overline{X} \cdot Z + W \cdot X \cdot Z + W \cdot \overline{X} \cdot \overline{Z}$$

W	X	Y	Z	F_9	F_{10}
0	0	0	0	1	1
0	0	0	1	X	0
0	0	1	0	0	1
0	0	1	1	1	0
0	1	0	0	1	X
0	1	0	1	X	0
0	1	1	0	0	1
0	1	1	1	0	0
1	0	0	0	1	0
1	0	0	1	X	0
1	0	1	0	1	1
1	0	1	1	0	X
1	1	0	0	0	X
1	1	0	1	1	0
1	1	1	0	0	1
1	1	1	1	1	0

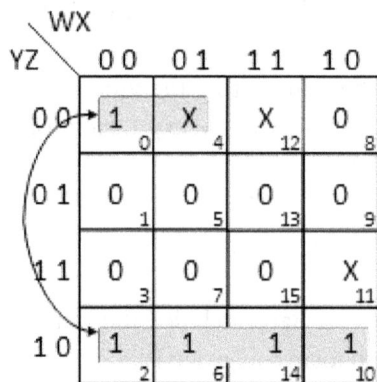

K-map for function

$$F_{10} = \overline{W} \cdot \overline{Z} + Y \cdot \overline{Z}$$

FIGURE 4.9. Two logic functions with Don't care terms.

4.3.1. MINIMIZING FUNCTIONS WITH DON'T CARES

To minimize a function with don't cares requires a little more care. The procedure is the same in the sense that we group together terms that are adjacent.

The additional care that needs to be exercised is with the don't care terms. For this, we need to understand the concept of *Cover*. The concept of cover says that we have to cover all the Min terms that have a logic High output, which means that we have to include every term that has a logic High output in the function. We do not have to include any terms that have a don't care output. We can, however, include those terms that have a don't care output if including the terms will help us in writing an expression that has a smaller cost. We will apply this concept on functions F_9 and F_{10}. Using the K-map drawn in Figure 4.9, we get Equation 4.18, which gives us the expression for function F_9.

$$F_9 = \overline{W} \bullet \overline{X} \bullet Z + \overline{W} \bullet \overline{Y} + W \bullet X \bullet Z + W \bullet \overline{X} \bullet \overline{Z} \qquad (4.18)$$

In writing the expression for function F_9, we have included the Min terms d_1 and d_5, which are don't care terms, as they help to form the group $\overline{W} \bullet \overline{Y}$. Without including these terms, the group would have been a group of only two terms and would have been $\overline{W} \bullet \overline{Y} \bullet \overline{Z}$. We have not included the Min term d_9, as this don't care term does not help in writing the function F_9 in any way that reduces its cost. Similarly, we can write the expression for the function F_{10}. This is given in Equation 4.19.

$$F_{10} = \overline{W} \bullet \overline{Z} + Y \bullet \overline{Z} \qquad (4.19)$$

In writing the expression for function F_{10}, we have included the Min term d_4, which is a don't care term, as it helps to form the group $\overline{W} \bullet \overline{Z}$. Without including these terms, the group would have been a group of only two terms and would have been $\overline{W} \bullet \overline{Z} \bullet \overline{Z}$. We have not included the Min terms d_{11} and d_{12}, as these don't care terms and do not help in writing the function F_{10} in any way that reduces its cost. Writing the expression in the POS expression also follows a similar procedure. Functions F_9 and F_{10} are written in POS expression in Equations 4.20 and 4.21.

$$F_9 = \left(W + \overline{Y} + Z\right) \bullet \left(W + \overline{X} + \overline{Y}\right) \bullet \left(\overline{W} + \overline{X} + Z\right) \bullet \left(\overline{W} + X + \overline{Z}\right) \quad (4.20)$$

$$F_{10} = Z \bullet \left(\overline{W} + Y\right) \qquad (4.21)$$

In writing the function F_9 in the POS expression, we have included the don't care Min term d_9 and have not included the don't care Min terms d_1 and d_5.

In writing the POS expression for function F_{10}, we find that we have included the Min terms d_{11} and d_{12}. In our example of the function F_9, we have used all the don't care terms in either the SOP or the POS expression. It does not have to be so. It is quite possible that for some functions one or more don't care terms are not included in either the SOP or the POS function.

Review Questions for Section 4.3

Question: Write the SOP expressions for the two, three-variable functions and the two four-variable, incompletely specified functions shown in Figure 4.10.

    Answer: The SOP expression for the K-map shown in Figure 4.8a is $\overline{W} \bullet \overline{Y} + W \bullet Y$.

The SOP expression for the K-map shown in Figure 4.8b is $\overline{Y} + W \bullet X$.

The SOP expression for the K-map shown in Figure 4.8c is $\overline{X} \bullet \overline{Z} + X \bullet \overline{Y}$.

The SOP expression for the K-map shown in Figure 4.8d is $\overline{X} \bullet Z + X \bullet \overline{Y} \bullet \overline{Z} + W \bullet Y \bullet Z$.

Question: Write the POS expressions for the two three-variable functions and the two four-variable, incompletely specified functions shown in Figure 4.10.

    Answer: The SOP expression for the K-map shown in Figure 4.10a4.8a is $\left(\overline{W} + Y\right) \bullet \left(W + \overline{Y}\right)$.

The POS expression for the K-map shown in Figure 4.10b is $\left(\overline{Y} + W\right) \bullet X$.

The POS expression for the K-map shown in Figure 4.10c is $\left(\overline{X} + \overline{Y}\right) \bullet \left(\overline{Z} + \overline{Y}\right) \bullet \left(X + \overline{Z}\right)$.

The POS expression for the K-map shown in Figure 4.10d is $\left(\overline{Y} + Z\right) \bullet \left(Z + X\right) \bullet \left(\overline{Z} + W + \overline{X}\right) \bullet \left(\overline{X} + Y + \overline{Z}\right)$.

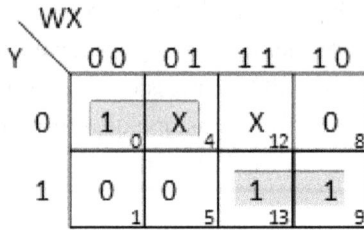

Figure 4.10a

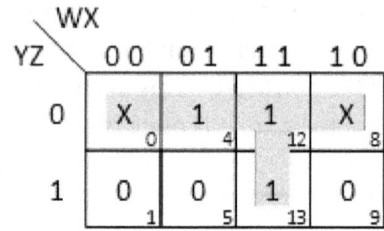

Figure 4.10b

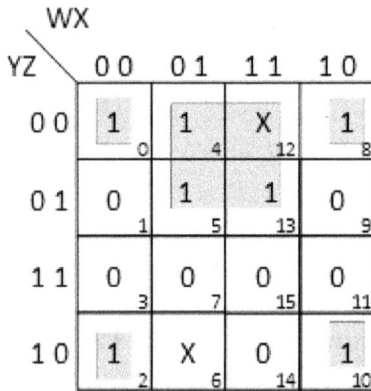

Figure 4.10c

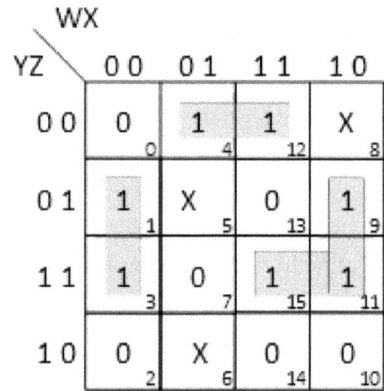

Figure 4.10d

FIGURE 4.10. Using a K-map to minimize functions with.

4.4.   USING THE TABULAR METHOD TO SIMPLIFY A LOGIC FUNCTION

In this section, we examine one computer-based method to build a two-level logic function. This is the Quine-McCluskey method. The justification for this method is based on the fact that a simple algorithm will give us a two-level logic expression in the AND-OR-INVERT format that is the minimum cost function. The expression that we get is as good as the one we can get using the K-map method. The K-map method is good as long as

	Quine – McCluskey	
Column 1	Column 2	Column 3

Column 1

Group 0 m₀ 0000 ✓
d₁ 0001 ✓
Group 1 m₄ 0100 ✓
m₈ 1000 ✓
m₃ 0011 ✓
Group 2 d₅ 0101 ✓
d₉ 1001 ✓
m₁₀ 1010 ✓
Group 3 m₁₃ 1101 ✓
Group 4 m₁₅ 1111 ✓

Column 2

m₀d₁ 000x ✓
Group 1 m₀m₄ 0x00 ✓
m₀m₈ x000 ✓
d₁m₃ 00x1 *
d₁d₅ 0x01 ✓
Group 2 d₁d₉ x001 ✓
m₄d₅ 010x ✓
m₈d₉ 100x ✓
m₈m₁₀ 10x0 *
d₅m₁₃ x101 ✓
Group 3 d₉m₁₃ 1x01 ✓
Group 4 m₁₃m₁₅ 11x1 *

Column 3

Group 2 m₀d₁ / m₄d₅ 0x0x *
m₀m₈ / d₁d₉ x00x *
Group 3 d₁d₅ / d₉m₁₃ xx01 *

FIGURE 4.11. Quine McCluskey tables for function F_9.

we have only a few variables in a function, but the K-map method is very difficult when we have either five or six variables. The Quine-McCluskey method does not care how many variables there are in the function. The Quine-McCluskey method is the same whether we have two variables or twenty-two variables. This method, developed in the mid-1950s, provides a very systematic procedure to get all the implicants and then extract the minimum set of essential implicants to cover the logic function. To understand how the method works, let us use the same example as in Figure 4.9, which was the function F_9. This is shown in Figure 4.11.

4.4.1. STARTING THE METHOD

As a first step, we write each of the Min terms, including the don't care terms, using their binary representation. In writing the Min terms, we have divided all the Min terms in several groups. The first group is Group 0; it consists of all the terms that have zero logic High values. The next group is

Group 1; it consists of all the terms that have exactly one logic High value. The next group is Group 2; it consists of all the terms that have only two logic High values. We continue this way until all the terms, including the don't care terms, have been entered. Notice that to separate the groups, we have drawn a line between them. This will aid us in forming pairs of terms that we will combine in the logic function. Not all the groups have to have entries. There may be groups that have no entries in them. We will make a note of such groups by leaving empty space for the group and drawing a separation line after the empty group. This way, we will remember that we have an empty group. For function F_9, we have started the QM method by entering all the Min terms divided in groups in Column 1 of Figure 4.11.

Combining Min terms: The goal of the method is to combine terms together according to the laws of Boolean Algebra to get to a minimum cost function. To do this, we compare the terms from two adjacent groups from Column 1 in Figure 4.11. We will compare terms from Group 0 with terms in Group 1; next, we will compare terms from Group 1 with the terms in Group 2. We will continue this process until all the groups have been compared. The procedure that we use to compare terms is to take one term from one group and compare it with all the terms in the next group. While comparing, we will use our Simplification Theorem a. Therefore, we are looking for two terms, one from each group, that differ from each other in only one position. For example, when we look at the term m_0 from Group 0 and d_1 from Group 1, we find that these two terms differ in only one position. The variable Z is 0 in the Min term m_0 and is 1 in the Min term d_1. All the other terms are identical. See Equation 4.22.

$$m_0 = \bar{W} \bullet \bar{X} \bullet \bar{Y} \bullet \bar{Z} \quad \text{and} \quad d_1 = \bar{W} \bullet \bar{X} \bullet \bar{Y} \bullet Z$$
$$\left(m_0 d_1\right) = \bar{W} \bullet \bar{X} \bullet \bar{Y} \bullet \left(\bar{Z} + Z\right) = \bar{W} \bullet \bar{X} \bullet \bar{Y} = 0 \bullet 0 \bullet 0 \bullet x \quad (4.22)$$

Since these two terms differ in only one position, they are combined together and entered in the next column in Group 1. We can similarly combine m_0 with the other two terms in Group 1 in Column 1 to form terms $m_0 m_4$ and $m_0 m_8$. These two terms are also entered in the second column as part of Group 1. As a matter of bookwork and as an aid to memory, we will put a check mark next to every term that we have combined with another term and transferred to the next column. Since we have transferred all the terms from Group 0 and Group 1 at least once, we have a check mark next to all the terms in Group 0 and Group 1.

This finishes comparing the terms in the first two groups. We will repeat the process by comparing every term in Group 1 (even though we have already transferred all the terms to the next column with previous combinations) with every term in Group 2 and combine them if possible. Combining the terms in Group 1 with the terms in Group 2, we find that we are able to form the following groups: d_1m_3, d_1d_5, d_1d_9, m_4d_5, m_8d_9, and m_8m_{10}. After the groupings are made, we will have transferred all the terms from Group 2 at least once to Column 2. To remember this, we have a check mark next to all the terms of Group 2 in Column 1. This completes the comparing process for Group 1 with Group 2.

We continue this way with the next pair of groups. The process is exactly the same. We compare the terms in two groups, find two that differ in only one position, and transfer them to the next column. When we transfer any term to the next column, we put a check mark next to the term. When we have compared all the groups in a column, we move to the next column and repeat the process.

To combine terms in the second column onward, we have to modify the process just a little. This time, when we are choosing terms to combine, we must compare only those terms that have don't cares in matching positions. In Group 1 and Group 2 in the second column, we will begin by comparing the first term, m_0d_1 in Group 1, with only the terms m_4d_5 and m_8d_9 in Group 2, as only these terms have the don't care in the same position. This comparison allows us to transfer the terms as groups of four to Group 2 in Column 3. The comparison process is exactly the same as before, and when we transfer any term to the next column, we put a check mark next to it.

Sometimes we have terms that do not combine with any other terms. In our example, this happens with terms d_1m_3, m_8m_{10}, and $m_{13}m_{15}$ in Column 2. Since we have not transferred these terms to the next column, we put a different mark next to them. We need to remember these terms for the next step in this method. All the terms that have not been transferred to the next column form the implicants of the function. From all the possible implicants, we have to find the prime implicants that cover the function. We will do this in the next section. As another example, let us use the QM method

Quine – McCluskey	
Column 1	Column 2

Group 0 m_0 0 0 0 0 *

Group 1 ─────────

Group 2 m_3 0 0 1 1 ✓
d_5 0 1 0 1 ✓
m_6 0 1 1 0 ✓
d_9 1 0 0 1 ✓
m_{10} 1 0 1 0 *
d_{12} 1 1 0 0 ✓

Group 3 m_7 0 1 1 1 ✓
m_{13} 1 1 0 1 ✓

Group 3 m_3m_7 0 x 1 1 *
d_5m_7 0 1 x 1 *
m_6m_7 0 1 1 x *
d_9m_{13} 1 x 0 1 *
d_1d_5 0 x 0 1 *
$d_{12}d_{13}$ 1 1 0 x *

FIGURE 4.12. Q-M tables for function F_{11}.

to determine all the implicants of the function F_{11} defined in Equation 4.23. The columns of the QM method for this function are shown in Figure 4.12.

$$F_{11} = \sum (m_0, m_3, m_6, m_7, m_{10}, m_{13}) + (d_5, d_9, d_{12}) \qquad (4.23)$$

When we enter the Min terms in the first column of the QM method, we find that there are no terms with only one logic High value, so Group 1 has no entries in it. That group is left blank. Next, when we combine terms from Group 2 with terms from Group 3, we find that the term m_{10} does not combine with any term, so it is marked with an *. We also mark term m_0 in Group 0 with an *, as this term also does not combine with any other term. In Column 2, we have only one group, so we are done combining terms and transferring them to the next column. All the implicants of the function are marked with an *.

4.4.2. DETERMINING THE PRIME IMPLICANTS

Now that we have the implicants of the function, we have to choose from them to determine the prime implicants that will cover the function. To do this, we begin with another table. This table is shown in Figure 4.13. The table in Figure 4.13 relates to the QM tables given in Figure 4.11. To begin the table that will enable us to choose the prime implicants, we start by listing all the Min terms that we have to cover the function across the top of the table. In doing this, we leave out all the don't care terms since we do not have to cover the don't care terms. We list these terms across the top of the table. To the left of the table, we list all the implicants that we have determined using the QM method. These are all the implicants that have an * next to them. To fill out the table, we put a check under every Min term that is covered by a particular implicant. So, for example, the first implicant is d_1m_3. This implicant covers the Min term m_3, so we have put a check under the Min term m_3 in the same row as the implicant d_1m_3. Similarly, the implicant $m_{13}m_{15}$ covers two Min terms, so in the row of this implicant, we have two check marks, one under m_{13} and the other under m_{15}. This way, we fill out the inside of the table.

To choose the prime implicants, we first search all the columns to see if any Min term is covered by only one implicant. We must choose this implicant as one of the prime implicants since this is the only implicant that covers the Min term under which there is only one check mark. Doing this in our table, we find that the Min term m_3 is covered by only one implicant.

Implicants	m_0	m_3	m_4	m_8	m_{10}	m_{13}	m_{15}	Prime Implicants
d_1m_3		✓						✓
m_8m_{10}				✓	✓			✓
$m_{13}m_{15}$						✓	✓	✓
m_0d_1 m_4d_5	✓		✓					✓
m_0m_8 d_1d_9	✓			✓				
d_1d_5 d_9m_{13}						✓		
Cover	✓	✓	✓	✓	✓	✓	✓	

FIGURE 4.13. Choosing the Prime Implicants.

This implicant is $d_1 m_3$; we will make this implicant a prime implicant. Now that we have made this implicant a prime implicant, we put a check mark on the right of the table in the row of this implicant. These check marks will tell us that the particular implicant is a prime implicant. By choosing this implicant as one of the prime implicants, we have covered the Min term m_3. When we cover a Min term, we put a check mark on the bottom of the column to indicate that this Min term is covered by the prime implicants chosen so far.

Continuing this way, we search other columns; next, we find that Min term m_{10} is covered by only one implicant. This implicant is $m_8 m_{10}$; we will make this implicant a prime implicant. Since this implicant covers two Min terms, we have covered both these Min terms when we choose this implicant as a prime implicant. Since the Min term m_8 is covered by this prime implicant, we do not need to cover it by any other implicant. Continuing this way, we find that m_4 is covered by only one implicant, $m_0 d_1 m_4 d_5$. This implicant must be chosen to be a prime implicant. We see that Min term m_{15} is also covered by only one implicant, $m_{13} m_{15}$; this implicant must be included as a prime implicant. When we choose the implicants that cover these Min terms, we have covered the entire function. We know that we have covered the entire function because we have check marks under every Min term in the table. This gives us a list of all the required prime implicants. For our example, the function is covered by four prime implicants. The minimum expression is shown in Equation 4.24.

$$F_9 = \overline{W} \bullet \overline{X} \bullet Z + \overline{W} \bullet \overline{Y} + W \bullet X \bullet Z + W \bullet \overline{X} \bullet \overline{Z} \qquad (4.24)$$

Compare Equation 4.24 with Equation 4.18. They are both the same, as they should be since both the expressions represent the minimum expression for function F_9.

As a second example, we use the implicants for function F_{11} determined in Figure 4.12. The details are shown in Figure 4.14. In Figure 4.14, across the top, we have listed all the Min terms that we have to cover. Down on the left side, we have listed all the implicants that we can use to cover the function. We first see that Min term m_0 and Min term m_{10} have not combined with any other term, so these terms will be chosen as a prime implicants. Next, we see that the Min terms m_3, m_6, and m_{13} are covered by only one

Implicants	m_0	m_3	m_6	m_7	m_{10}	m_{13}	Prime Implicants
m_0	✓						✓
m_{10}					✓		✓
m_3m_7		✓		✓			✓
d_5m_7				✓			
m_6m_7			✓	✓			✓
d_9m_{13}						✓	✓
d_1d_5							
$d_{12}d_{13}$							
Cover	✓	✓	✓	✓	✓	✓	

FIGURE 4.14. Choosing the Prime Implicants.

implicant each; these implicants, m_3m_7, m_6m_7, and d_9m_{13}, have to be chosen as the prime implicants. Choosing the five prime implicants covers the entire function.

The QM method is simple, but it does require a lot of care. It will always give us a minimum two-level AND-OR-INVERT function.

Review Question for Section 4.4

Question: Use the QM method to determine the minimum expression for the four functions given in Figure 4.10. The four QM tables to first select the implicants and then to determine the prime implicants are all given in Figure 4.15.

Answer: The SOP expression for the QM table shown in Figure 4.15a is $\bar{W} \bullet \bar{Y} + W \bullet Y$.

The SOP expression for the QM table shown in Figure 4.15b is $\bar{Y} + W \bullet X$.

The SOP expression for the QM table shown in Figure 4.15c is $\bar{X} \bullet \bar{Z} + X \bullet \bar{Y}$.

The SOP expression for the QM table shown in Figure 4.15d is $\bar{X} \bullet Z + X \bullet \bar{Y} \bullet \bar{Z} + W \bullet Y \bullet Z$.

4.5. SOME SIMPLE APPLICATIONS OF LOGIC CIRCUITS AND LOGIC FUNCTIONS

The logic functions that we have used and the truth tables that we have derived in this chapter and the previous chapters—where do they come from? How do we use the logic function and the truth table? To answer this question, let us examine a very simple car security system. To understand the entire process, let us first define what we want in the car security system. We want the security system to give a continuous blast on the car horn when an alarm condition is triggered. In addition to this, we will also have an activation switch. This switch will arm the security system or will completely disconnect the car horn from the security system. In addition to this, let us assume that the car has two sensors on it. These two sensors make up the security system. The first sensor is a vibration sensor. This will activate when the car is shaken, which can happen when someone tries to tow the car away. The second sensor is a door sensor. This sensor triggers when someone tries to open the car door. All these statements give us the conditions that we have to use to make up a car alarm system.

When master switch M is active, it will represent a logic High or a 1. When this switch is deactivated, it will represent a logic Low or a 0.

When the vibration sensor V detects vibration, it will put out a logic High or a 1. When no vibration is sensed, this sensor will put out a 0.

Figure 4.15a

Column 1	Column 2
m_0 000 ✓	m_0d_2 0 x 0 *
d_2 010 ✓	d_2d_6 x 1 0 *
m_5 101 ✓	m_5m_7 1 x 1 *
d_6 110 ✓	
m_7 111 ✓	

Implicants	m_0	m_5	m_7	Prime Implicants
m_0d_2	✓			✓
d_2d_6				
m_5m_7		✓	✓	✓
Cover	✓	✓	✓	

Figure 4.15a

Figure 4.15b

Column 1	Column 2	Column 3
d_0 000 ✓	d_0m_2 0 x 0 ✓	d_0m_2
m_2 010 ✓	d_0d_4 x 0 0 ✓	d_4m_6 x x 0 *
d_4 100 ✓	m_2m_6 x 1 0 ✓	
m_6 110 ✓	d_4m_6 1 x 0 ✓	
m_7 111 ✓	m_6m_7 1 1 x *	

Implicants	m_2	m_6	m_7	Prime Implicants
m_6m_7		✓	✓	✓
d_0m_2 d_4m_6	✓			✓
Cover	✓	✓	✓	

Figure 4.15b

Figure 4.15c

Column 1	Column 2	Column 3
m_0 0000 ✓	m_0m_2 00x0 ✓	m_0m_2 m_4m_6 0xx0 *
m_2 0010 ✓	m_0m_4 0x00 ✓	m_0m_2 m_8m_{10} x0x0 *
m_4 0100 ✓	m_0m_8 x000 ✓	m_0m_4 m_8m_{12} xx00 *
m_8 1000 ✓	m_2d_6 0x10 ✓	m_4m_5 $d_{12}m_{13}$ x10x *
m_5 0101 ✓	m_2m_{10} x010 ✓	
d_6 0110 ✓	m_4m_5 010x ✓	
m_{10} 1010 ✓	m_4m_6 01x0 ✓	
d_{12} 1100 ✓	m_4d_{12} x100 ✓	
m_{13} 1101 ✓	m_8m_{10} 10x0 ✓	
	m_8d_{12} 1x00 ✓	
	$d_{12}m_{13}$ 110x ✓	
	m_5m_{13} 1x01 ✓	

Implicants	m_0	m_2	m_4	m_5	m_8	m_{10}	m_{13}	Prime Implicants
m_0m_2 m_4d_6	✓	✓	✓					
m_0m_2 m_8m_{10}	✓	✓			✓	✓		
m_0m_4 m_8d_{12}	✓		✓		✓			
m_4m_5 $d_{12}m_{13}$			✓	✓			✓	
Cover	✓	✓	✓	✓	✓	✓	✓	

Figure 4.15c

Figure 4.15d

Column 1	Column 2	Column 3
m_1 0001 ✓	m_1m_3 00x1 ✓	m_1m_3 m_9m_{11} xx01 *
m_4 0100 ✓	m_1d_5 0x01 *	
m_8 1000 ✓	m_1m_9 x001 *	
m_3 0011 ✓	m_4d_5 010x *	
d_5 0101 ✓	m_4d_6 01x0 *	
d_6 0110 ✓	m_4m_{12} x100 *	
m_9 1001 ✓	d_8m_9 100x *	
m_{12} 1100 ✓	d_8m_{12} 1x00 *	
m_{11} 1011 ✓	m_3m_{11} x011 ✓	
m_{15} 1111 ✓	m_9m_{11} 10x1 ✓	
	$d_{11}m_{15}$ 1x11 *	

Implicants	m_1	m_3	m_4	m_9	m_{11}	m_{12}	m_{15}	Prime Implicants
m_1m_3 m_9m_{11}	✓	✓		✓	✓			✓
m_1d_5	✓							
m_4d_5			✓					
m_4d_6			✓					
m_4m_{12}			✓			✓		✓
d_8m_9				✓				
$m_{11}m_{15}$					✓		✓	✓
Cover	✓	✓	✓	✓	✓	✓	✓	

Figure 4.15d

FIGURE 4.15. Using QM method to minimize the given functions

When the door sensor D detects that a door is opened, it will put out a logic High or a 1. When all the doors are locked, the door sensor will put out a 0.

We can make a statement of the security system as follows: There are three conditions when the car horn should sound to alert the owner. They are as follows:

a. The master switch M is activated, the vibration sensor V is High, and the door sensor D is Low.

b. The master switch M is activated, the vibration sensor V is Low, and the door sensor D is High.

c. The master switch M is activated, the vibration sensor V is High and the door sensor D is High.

When any one or more of these conditions are met, the car horn will sound. Given the above description of the car security system, we can first define a function that will turn the car horn, as shown in Equation 4.25.

$$H = M \bullet V \bullet \bar{D} + M \bullet \bar{V} \bullet D + M \bullet V \bullet D \qquad (4.25)$$

A logic function is simply a concise way of writing the condition that we want to detect. Now that we have the logic function, we can talk about building the truth table. To build the truth table, we first list all the possible combinations of the input variables. In our example, we have three variables that are used to build the logic function. Each of the variables will have either a value of 0 or a value of 1. With three variables, our truth table will have eight rows, as shown in Figure 4.16. In Figure 4.16, we also see the K-map for the function and the logic equation for the function.

M	V	D	H
0	0	0	0
0	0	1	0
0	1	0	0
0	1	1	0
1	0	0	0
1	0	1	1
1	1	0	1
1	1	1	1

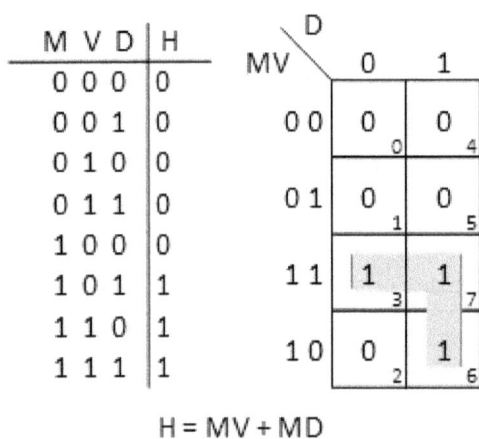

$H = MV + MD$

FIGURE 4.16. A simple car alarm security system.

4.6. CHAPTER PROBLEMS

4.6.1. Using the K-map method, determine the minimum cost SOP expression for the following functions.

4.6.1.1. $Q_1 = A \bullet \bar{B} \bullet C + \bar{A} \bullet B \bullet C + A \bullet \bar{B} \bullet \bar{C} + \bar{A} \bullet B \bullet \bar{C}$

4.6.1.2. $Q_2 = A \bullet B \bullet C + \bar{A} \bullet \bar{B} \bullet C + A \bullet \bar{B} \bullet \bar{C} + \bar{A} \bullet B \bullet \bar{C}$

4.6.1.3. $Q_3 = A \bullet \bar{B} \bullet C + \bar{A} \bullet \bar{B} \bullet C + A \bullet \bar{B} \bullet \bar{C} + \bar{A} \bullet B \bullet C$

4.6.1.4. $Q_4 = A \bullet \bar{B} \bullet C + \bar{A} \bullet \bar{B} \bullet C + A \bullet \bar{B} \bullet \bar{C} + \bar{A} \bullet \bar{B} \bullet \bar{C}$

4.6.2. Using the K-map method, determine the minimum cost POS expression for all the functions in Problem 4.6.1.

4.6.3. Using the K-map method, determine the minimum cost SOP expression for the following functions.

4.6.3.1. $Q_5 = \bar{A} \bullet \bar{B} \bullet \bar{C} \bullet \bar{D} + \bar{A} \bullet B \bullet \bar{C} \bullet \bar{D} + A \bullet B \bullet C \bullet \bar{D}$
$+ A \bullet \bar{B} \bullet C \bullet D + \bar{A} \bullet \bar{B} \bullet C \bullet D + A \bullet B \bullet C \bullet D$

4.6.3.2. $Q_6 = \bar{A} \bullet B \bullet \bar{C} \bullet \bar{D} + \bar{A} \bullet B \bullet \bar{C} \bullet D + A \bullet \bar{B} \bullet \bar{C} \bullet D$
$+ A \bullet \bar{B} \bullet C \bullet D + \bar{A} \bullet \bar{B} \bullet C \bullet \bar{D} + \bar{A} \bullet B \bullet C \bullet \bar{D}$

4.6.3.3. $Q_7 = \bar{A} \bullet \bar{B} \bullet \bar{C} \bullet D + \bar{A} \bullet B \bullet \bar{C} \bullet D + A \bullet B \bullet \bar{C} \bullet D$
$+ A \bullet \bar{B} \bullet C \bullet D + A \bullet \bar{B} \bullet C \bullet \bar{D} + A \bullet B \bullet C \bullet D$

4.6.3.4. $Q_8 = \bar{A} \bullet \bar{B} \bullet \bar{C} \bullet \bar{D} + A \bullet B \bullet \bar{C} \bullet \bar{D} + A \bullet \bar{B} \bullet \bar{C} \bullet \bar{D}$
$+ \bar{A} \bullet \bar{B} \bullet C \bullet D + \bar{A} \bullet B \bullet C \bullet \bar{D} + A \bullet B \bullet C \bullet \bar{D}$

4.6.4. Using the K-map method, determine the minimum cost POS expression for all the functions in Problem 4.6.3.

4.6.5. Using the K-map method, determine the minimum cost SOP expression for the following incompletely specified functions.

4.6.5.1. $Q_9 = \sum (m_0, m_3, m_4, m_7, m_{12}, m_{15}) + (d_2, d_9, d_{13})$

4.6.5.2. $Q_{10} = \sum (m_0, m_2, m_4, m_6, m_{12}, m_{13}) + (d_5, d_8, d_{14})$

4.6.5.3. $Q_{11} = \sum (m_1, m_6, m_7, m_9, m_{11}, m_{13}) + (d_2, d_8, d_{12})$

4.6.5.4. $Q_{12} = \sum (m_1, m_5, m_8, m_9, m_{11}, m_{14}) + (d_2, d_8, d_{13})$

4.6.6. Using the K-map method, determine the minimum cost POS expression for all the functions in Problem 4.6.5.

4.6.7. Using the K-map method, determine the minimum cost SOP expression for the following incompletely specified functions.

4.6.7.1. $Q_{13} = \prod(M_0, M_3, M_4, M_7, M_{12}, M_{15}) + (D_2, D_9, D_{13})$

4.6.7.2. $Q_{14} = \prod(M_0, M_2, M_4, M_6, M_{12}, M_{13}) + (D_5, D_8, D_{14})$

4.6.7.3. $Q_{15} = \prod(M_1, M_6, M_7, M_9, M_{11}, M_{13}) + (D_2, D_8, D_{12})$

4.6.7.4. $Q_{16} = \prod(M_1, M_5, M_8, M_9, M_{11}, M_{14}) + (D_2, D_8, D_{13})$

4.6.8. Using the K-map method, determine the minimum cost POS expression for all the functions in Problem 4.6.7.

4.6.9. Using the QM method, determine the minimum cost SOP expression for all the functions in Problem 4.6.1.

4.6.10. Using the QM method, determine the minimum cost SOP expression for all the functions in Problem 4.6.3.

4.6.11. Using the QM method, determine the minimum cost SOP expression for all the functions in Problem 4.6.5.

4.6.12. Using the QM method, determine the minimum cost SOP expression for all the functions in Problem 4.6.7.

4.6.13. Use a four-variable K-map for this problem. Fill it with 1's and 0's to define a function. This function has to satisfy the following conditions:

4.6.13.1. The minimum SOP and the POS have the same number of terms in them.

4.6.13.2. The minimum SOP has fewer terms than the minimum POS expression.

4.6.13.3. The minimum SOP has fewer terms than the minimum POS expression.

4.6.13.4. There are at least two groups or eight terms in the SOP expression.

4.6.13.5. There are at least two groups or eight terms in the POS expression.

4.6.14. Consider a five-input Boolean function that is true when exactly two of its inputs are High. For this function:

4.6.14.1. Build a truth table.

4.6.14.2. Write the canonical SOP expression for this function.

4.6.14.3. Write the canonical POS expression for this function.

4.6.14.4. Use the QM method to determine the minimum SOP expression.

4.6.15. Consider a four-input Boolean function that is true when an odd number of its inputs are High. For this function:

4.6.15.1. Build a truth table.

4.6.15.2. Write the canonical SOP expression for this function.

4.6.15.3. Write the canonical POS expression for this function.

4.6.15.4. Use the K-map method to determine the minimum SOP expression.

4.6.16. Consider a five-input Boolean function that is true when an even number of its inputs are High. For this function:

4.6.16.1. Build a truth table.

4.6.16.2. Write the canonical SOP expression for this function.

4.6.16.3. Write the canonical POS expression for this function.

4.6.16.4. Use the QM method to determine the minimum SOP expression.

4.7. APPENDIX 4.A: THE GRAY CODE

Earlier in this chapter, we used the Gray code to list the possible headings of the rows and the columns of the K-map. We use the Gray code because this code has a special property. If you examine the Gray code, you will see this property right away. The property is as follows: when you advance from one index to the next, there is a change in only a single bit. This property is not true in the regular binary code. Examine the Gray code number wheel shown in Figure 4.17.

We expect this exact same property in adjacent squares on the K-map so that we can combine the squares and form larger groups. For this reason, we use the Gray code to head the rows and the columns of the K-map. Since

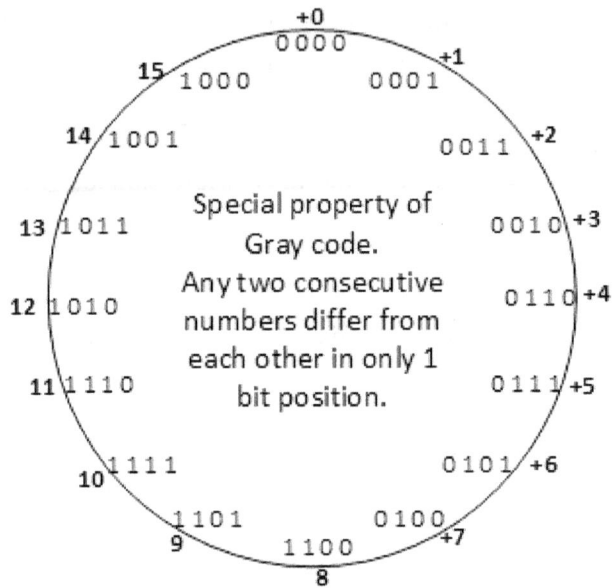

FIGURE 4.17. Representing numbers in Gray code.

we are going to use the Gray code, we need to be able to write the Gray code of different sizes. We can use a simple algorithm to generate the Gray code. This algorithm is discussed in Figure 4.18.

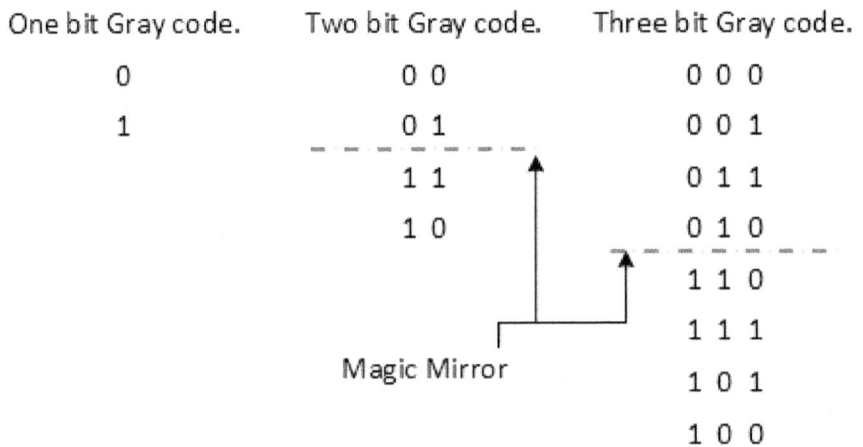

One bit Gray code.	Two bit Gray code.	Three bit Gray code.
0	0 0	0 0 0
1	0 1	0 0 1
	1 1	0 1 1
	1 0	0 1 0
		1 1 0
		1 1 1
		1 0 1
		1 0 0

Magic Mirror

FIGURE 4.18. Algorithm for generating Gray code.

INTRODUCTION TO DIGITAL LOGIC DESIGN

We begin with a one-bit Gray code. This is a simple Gray code, as it is only 0 and a 1, as shown in Figure 4.18. From the one-bit Gray code, we get a two-bit Gray code as follows. First, we add the extra bit by placing a 0 in the most significant position of the one-bit Gray code. We then get a "magic mirror" and put it under this code. This mirror exactly reflects the Gray code that we had (that is the one-bit Gray code), but it complements the added bits that were all 0's. This way, we now have a two-bit Gray code.

Next, to get a Gray code of one higher dimension, we follow the same procedure. First, we add the extra bit by placing a 0 in the most significant position of the Gray code that we have. We then get a "magic mirror" and put it under this code. This mirror exactly reflects the Gray code that we had, but it complements the added bits that were all zeros. This way, we now have a Gray code that is one dimension higher than the one that we had. Verify this with the three-bit Gray code given in Figure 4.18 and the four-bit Gray code given in Figure 4.17.

5. ARITHMETIC IN BINARY

5.0.   INTRODUCTION

In Chapter 1, we saw how we can represent numbers using the binary number system. At that time, we only looked at the magnitude of the numbers. We did not pay any attention to negative numbers. In this chapter, we will investigate how negative binary numbers are represented. After we have represented both the positive and the negative numbers, we will want to do arithmetic with these numbers. To perform both the addition and subtraction, we will examine how an Arithmetic Logic Unit (ALU) is designed and built. The commercially available ALUs have circuits that help to speed the arithmetic.

5.1.   REPRESENTING POSITIVE AND NEGATIVE NUMBERS IN BINARY

In the number system, there are infinitely many positive and negative numbers. When we are using paper and pencil, we have no difficulty in writing any number. When we are doing arithmetic on a machine like a calculator or a computer, then we have only a finite space to write these numbers. For example, on your calculator, you have a display that has only nine digits, so the largest number that you can display is a nine-digit number. If you have a number that requires more than nine digits, then you have to use the exponential notation. In a modern computer, the numbers are represented using the binary number system. In the computer, the size of any number is limited to thirty-two binary digits, also known as bits. Using these thirty-two bits, we have to represent both positive and negative numbers. To understand how we represent both positive and negative numbers, we will use the example of numbers represented using only four bits.

Using four-bit representation, we can represent sixteen different numbers. Of these sixteen numbers, about half the numbers will be positive and about half the numbers will be negative. We will also look at three different methods to represent the numbers. Each of the three methods represents the positive numbers in the same way, but all three methods use a different representation for the negative numbers.

5.1.1. THE SIGN MAGNITUDE REPRESENTATION

To represent the numbers in the sign magnitude representation, we use the most significant digit to represent the sign and the remaining bits to represent the magnitude of the number. When the sign bit is zero, we have a positive number, and when the sign bit is a one, we have a negative number, as shown in Figure 5.1. The number wheel shows us all the numbers using four bits; Figure 5.1 shows us how we represent the numbers using the sign magnitude representation in the form of a number wheel. There are eight positive numbers and eight negative numbers. Using the sign magnitude representation, we see that there are two different representations for the number zero. Even though positive zero and negative zero do not have any difference, these two representations of zero are available when we use the sign magnitude representation.

The reason we represent numbers in any particular way is to do arithmetic with the numbers. Here, we will examine how we do arithmetic using the sign magnitude representation. Adding two positive numbers or two negative numbers is rather simple; we just add the magnitudes and copy

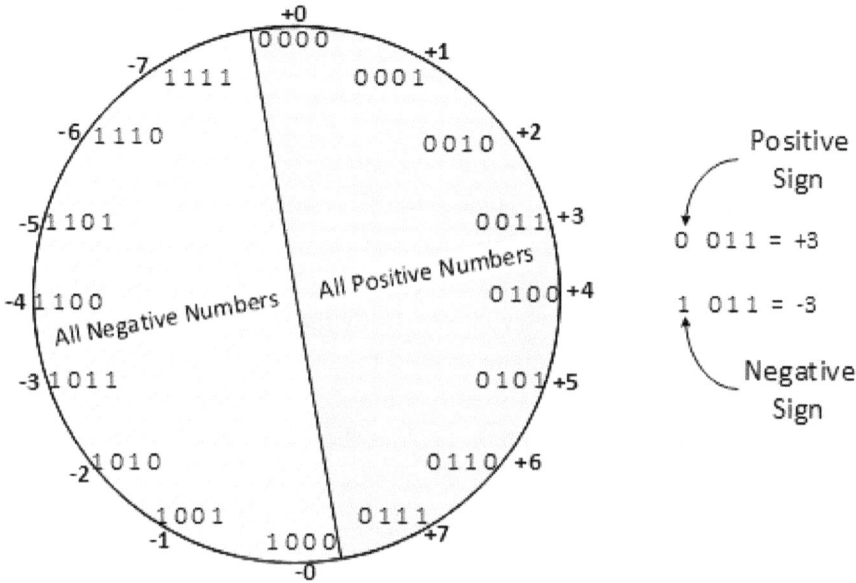

FIGURE 5.1. Representing positive and negative numbers using sign magnitude representation.

the sign from the two operands. Adding or subtracting when one number is positive and the other is negative, we run into a situation that is rather difficult to solve. To describe this problem, consider that we wish to add a positive number to a negative number. This requires that we subtract the smaller number from the larger number and then copy the sign of the larger number.

To complete this operation, we would need a subtractor to do the required arithmetic, and to determine the sign of the result, we need a comparator. This is required since the sign of the result will be the same sign as the sign of the larger magnitude number. This is what makes arithmetic with the sign magnitude representation very cumbersome. It is due to this complexity that the hardware designers have decided not to use the sign magnitude representation to perform arithmetic in binary.

5.1.2. 1'S COMPLEMENT REPRESENTATION

The 1's complement representation represents the positive numbers in the same way as we represent the positive numbers in the sign magnitude representation. The difference is in how we represent the negative numbers. First, we determine how to calculate the 1's complement of any number. To obtain the 1's complement of any number, we use the following procedure. If the number is a four-digit number, then we begin by writing the largest number using four digits. Since we are using the binary number system, the largest number using four digits will be **1111**. Next, to obtain the 1's complement representation of any three-digit number, we subtract the three-digit number from the largest number, 1111, as shown in Equation 5.1, where we determine the 1's complement of the number three.

$$
\begin{array}{ll}
1111 & \\
0011 & +3 \\
1100 & 1\text{'s complement of 3 or -3}
\end{array}
\qquad (5.1)
$$

The result of the subtraction is the 1's complement of that positive number. This is equal to the negative representation of that number in 1's complement representation. The number wheel of 1's complement representation of numbers for four-digit numbers is shown in Figure 5.2. Notice that all the negative numbers still have a 1 as the most significant digit and all the positive numbers have the 0 as the most significant digit. With the 1's complement representation, we still have two different representations of the number zero. The advantage that this representation provides us is that we can perform subtraction using addition, as shown in Equation 5.2.

$$
A - B = A + (-B) = A + (1\text{'s complement of B})
\qquad (5.2)
$$

To subtract the number B from the number A, we determine the 1's complement of the number B and then add it to the number A to get the result. Thus, when we use the 1's complement representation, we do not need a separate subtractor or a comparator to do the required arithmetic. Even though we do not need a separate subtractor, addition or subtraction

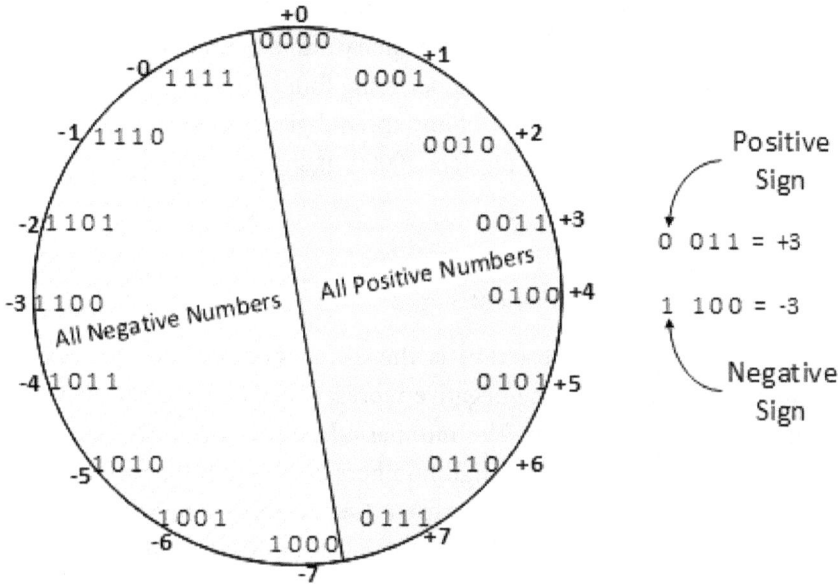

FIGURE 5.2. Representing positive and negative numbers using 1's complement notation.

is still complicated because we have two different representations of the number zero. To get around this difficulty, we move on to the 2's complement representation.

There is also an alternative way to obtain the 1's complement of a number. This alternative method is probably easier than the method outlined above. In this second procedure, we complement all the digits in the number to get the 1's complement of the number.

5.1.3. 2'S COMPLEMENT REPRESENTATION

The 2's complement representation represents the positive numbers in the same way as we represent the positive numbers in the sign magnitude representation and the 1's complement representation. The difference is in how we represent the negative numbers. To obtain the 2's complement of any number, we use the following procedure. If the number is a four-digit number, then we begin by writing a five-digit number, which is 1 followed by four zeros or **10000**. Since we are using the binary number system, this number is just one more than the largest number using four digits. Next, to obtain the 2's complement representation of any three-digit number, we subtract the three-digit number from the number 10000, as shown in Equation 5.3.

$$
\begin{array}{ll}
1\,0\,0\,0\,0 & \\
\;\,0\,0\,1\,1 & +3 \\ \hline
\;\,1\,1\,0\,1 & \text{2's complement of 3 or -3}
\end{array}
\qquad (5.3)
$$

The result of the subtraction is the 2's complement of that positive number. This is equal to the negative representation of that number in 2's complement representation. The number wheel of 2's complement representation of numbers for four-digit numbers is shown in Figure 5.3. Notice that all the negative numbers still have a 1 as the most significant digit and all the positive numbers have the 0 as the most significant digit. With the 2's complement representation, we now have only one representation for zero.

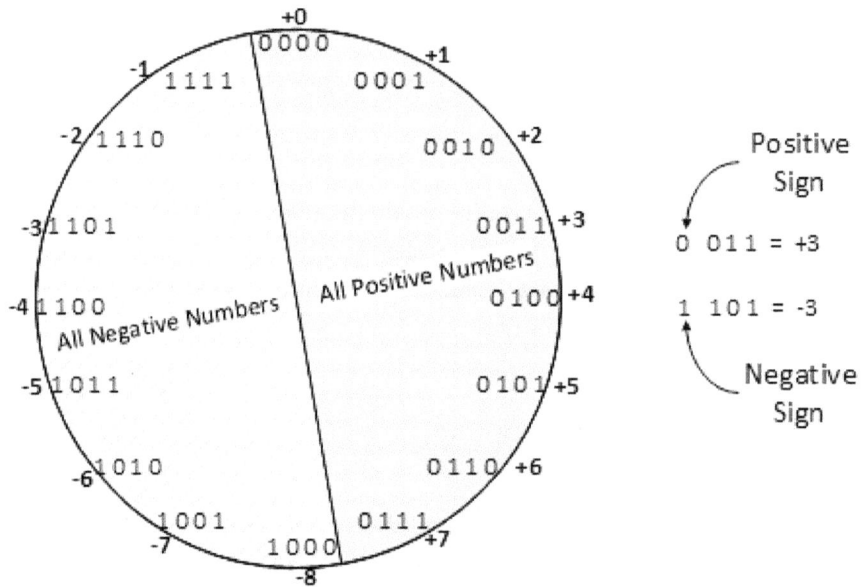

FIGURE 5.3. Representing positive and negative numbers using 2's complement notation.

The advantage that this representation provides us is that we can perform subtraction using addition, as shown in Equation 5.4.

$$A - B = A + (-B) = A + (2\text{'s complement of } B) \qquad (5.4)$$

To subtract the number B from the number A, we determine the 2's complement of the number B and then add it to the number A to get the result. Thus, when we use the 2's complement representation, we do not need a separate subtractor to do the required arithmetic. Now with this representation, we are able to perform the subtraction with no difficulty.

There is also an alternative way to obtain the 2's complement of a number. This is probably easier than the method outlined above. Here is the procedure: Step one: Complement all the digits in the number. This is the 1's complement of the number. Step two: Add one to the result in step one to get the 2's complement.

Review Questions for Section 5.1

Question: Convert the following numbers from base 10 to numbers in base 2 using the sign magnitude representation for negative numbers. Assume all numbers are ten-bit numbers in base 2. 8; −143; 52; −19; −33.

   Answer: 8 → 0000001000; −143 → 1010001111; 52 → 0000110100; −19 → 1000010011; −33 → 100100001.

Question: Convert the following numbers from base 10 to numbers in base 2 using the 1's complement representation for negative numbers. Assume all numbers are ten-bit numbers in base 2. 8; −143; 52; −19; −33.

   Answer: 8 → 0000001000; −143 → 1101110000; 52 → 0000110100; −19 → 1111101100; −33 → 111011110.

Question: Convert the following numbers from base 10 to numbers in base 2 using the 2's complement representation for negative numbers. Assume all numbers are ten-bit numbers in base 2. 8; −143; 52; −19; −33.

   Answer: 8 → 0000001000; −143 → 1101110001; 52 → 0000110100; −19 → 1111101101; −33 → 111011111.

5.2. BINARY ADDITION

In this section, we will see how addition and subtraction are performed in all three representations of binary numbers. After we have seen how the addition is done, we will be able to see why the engineers have chosen to use the 2's complement representation to represent numbers in a computer system. For this, we will continue to use our four-bit numbers that we have available in the number wheels in Figures 5.1, 5.2, and 5.3.

5.2.1. ADDING BINARY NUMBERS

This should remind you of what you did when you were in second grade in elementary school. We are going to see how simple addition is done when we are using binary numbers. Just as we did in elementary school, we will build addition tables and then use the addition tables to complete the required addition. Look at Figure 5.4 for simple addition of binary digits.

For adding two binary digits, we have four different options. All four options are shown in Figure 5.4 in the top row. The first three sums are relatively simple to understand and are self-explanatory. In the last one, where we add $(1)_2$ to $(1)_2$, we get a result that is a 0 with a carryover to the next digit, with the carryover being 1. The following couple of examples will make this clear.

Examples of Adding.

```
   1 0 1 1        1 1 0 0        1 0 1 1
 + 0 1 1 0      + 1 0 1 1      + 1 1 1 0
 ¹‾0‾0‾0‾1‾      ¹‾0‾1‾1‾1‾      ¹‾1‾0‾0‾1‾
```

Examine the three examples of adding four-bit numbers given above. To complete the required addition, we use the addition tables given in Figure 5.4 bit by bit, starting from the least significant digit. For example, in the first addition, we begin by adding (1) to (0) so we get a result of (1) with no carryover. To add the next bit, we add (1) to (1) so we get a result of (0) with a (1) carryover. In the third digit, we add a (0) to (1), and to the

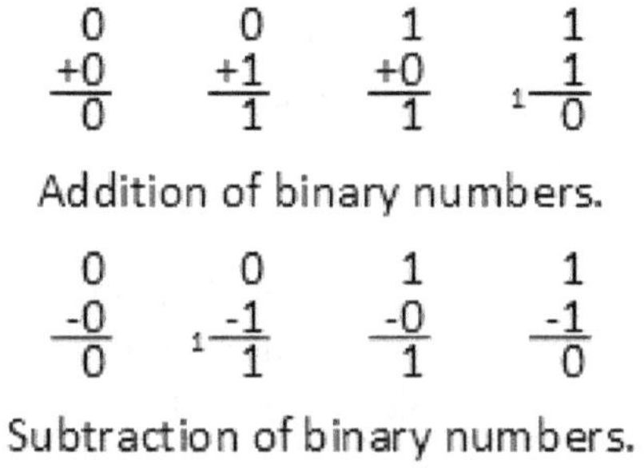

```
  0        0        1        1
 +0       +1       +0      ₁ 1
 ──       ──       ──      ──
  0        1        1        0
```

Addition of binary numbers.

```
  0        0        1        1
 -0      ₁ -1       -0       -1
 ──       ──       ──       ──
  0        1        1        0
```

Subtraction of binary numbers.

FIGURE 5.4. Addition and Subtraction tables for binary numbers.

result, we have to add the carryover from the previous bits, which is a (1). This gives us a result of (0) with a carryover of (1). In the forth digit, we add a (1) to (0), and to the result, we have to add the carryover from the previous bit, which is a (1). This gives us a result of (0) with a carryover of (1). This completes the addition. In the same way, you can verify the other two additions.

Examples of Subtraction.

```
  1011      1100       1011
 -0110     -1011      -1110
 ─────     ─────     ─────
  0101      0001     ₁ 1101
```

Examine the three examples of subtracting four-bit numbers given above. To complete the required subtraction, we use the subtraction tables given in Figure 5.4. In the first example, we begin by subtracting (0) from (1) so we get a result of (1) with no borrow. Next, we subtract (1) from (1) so we get a result of (0) with no borrow. In the third digit, we subtract a (1) from (0); this gives us a result of (1) with a borrow of (1). Since we have borrowed from the next digit, in the fifth position, the number in the forth position changes from a (1) to a (0). With the change, we are subtracting a (0) from (0). Remember we borrowed a one to complete the subtraction of the previous digit; this gives us a result of (0) with no borrow. This completes the subtraction. In the same way, you can verify the other two subtractions.

5.2.2. ARITHMETIC USING SIGN MAGNITUDE REPRESENTATION

To begin, let us look at some examples of adding and subtracting with the digits 2 and 4 when the numbers are written in the sign magnitude representation. There are four possible options; these options are shown in Figure 5.5.

In the two additions on the left in Figure 5.5, the sign of both the numbers is the same, so the result of addition is simply the addition of the magnitude portion and then copying the sign from the two numbers to get the sign of the result. The two additions on the right are more complex additions. They are complex because the sign of the two numbers is different. In the top sum, we want to add a negative 4 to a positive 2. To do this, we have to subtract 4 from 2. We know that the result will be a negative result. To get the result, we have subtracted 2 (the smaller number) from 4 (the larger number) and assigned the sign of the result to be the same as the sign of the larger number. With this, we get the result as −2 for our final answer. The sum on the bottom right is performed in a similar manner. We want to add a negative 2 to a positive 4, so we have to subtract 2 from 4. When we do this, we know the result will be a positive number. To complete the required addition, we subtract the smaller number (2) from the larger number (4) and get the result 2. To this result, we attach a positive sign, the sign of the larger number, to get the result of addition.

Sign magnitude arithmetic is complicated because we need both an adder and a subtractor. The adder is used when the signs of the two digits are the same, and the subtractor is used when the signs of the two digits are different. Then we also need a comparator to decide what the sign of the result

$$
\begin{array}{rl}
2 & 0010 \\
+4 & +0100 \\
\hline
6 & 0110
\end{array}
\qquad
\begin{array}{rl}
2 & 0010 \\
-4 & +1100 \\
\hline
-2 & 1010
\end{array}
$$

$$
\begin{array}{rl}
-2 & 1010 \\
-4 & +1100 \\
\hline
-6 & 1110
\end{array}
\qquad
\begin{array}{rl}
-2 & 1010 \\
4 & +0100 \\
\hline
2 & 0010
\end{array}
$$

FIGURE 5.5. Adding positive and negative numbers using sign magnitude representation.

should be when the two digits have opposite signs. For these reasons, the sign magnitude representation is not used for arithmetic and to represent numbers in a computer system.

5.2.3. ARITHMETIC USING 1'S COMPLEMENT REPRESENTATION

To begin, let us look at some examples of adding and subtracting with the digits 2 and 4 when the numbers are written in the 1's complement representation. There are four possible options; these options are shown in Figure 5.6.

This time, adding the two positive numbers gives us the same result that we got in the previous example, and that is what we expect. This should not be too difficult to understand since we represent the positive numbers in the same way in all three representations of numbers.

Adding two negative numbers, we see a major difference. This is the concept of the *end around carry*. When we add $(-2 \rightarrow 1101)$ to $(-4 \rightarrow 1011)$, we get a carry out. We loop this carry around to the least significant digit and add it to the previous result. This gives us the result $(-6 \rightarrow 1001)$, which is the correct result. The end around carry also occurs when we add (-2) and (4) to each other. Again, adding the end around carry to the least significant digit gives us the correct answer. In the $\{(+2) +(-4)\}$ addition, there is no end around carry, so the result of the addition is the expected answer.

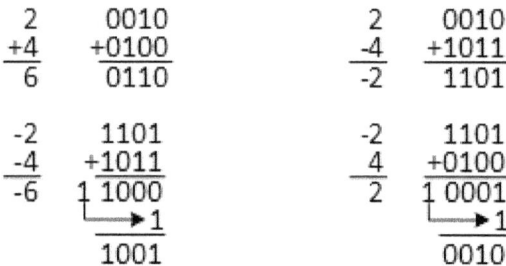

```
  2      0010           2      0010
 +4     +0100          -4     +1011
 ──      ────          ──      ────
  6      0110          -2      1101

 -2      1101          -2      1101
 -4     +1011           4     +0100
 ──     ──────         ──     ──────
 -6    1 1000           2    1 0001
          └──► 1                └──► 1
         ──────               ──────
          1001                 0010
```

FIGURE 5.6. Adding positive and negative numbers using 1's complement representation.

Why does the end around carry scheme work? It works because every time we have an end around carry we have crossed the origin of the number wheel. Remember that in the 1's complement, we have two representations for the number zero. To avoid counting the zero twice, we add the end around carry that in effect eliminates one of the zeros. We can show when the end around carry occurs and what is happening if we look at a general addition, as shown in Equation 5.5.

$$M - N = M + \left(\underbrace{2^n - 1 - N}_{\substack{\text{Obtaining 1's} \\ \text{complement of N}}} \right) = (M - N) + 2^n - 1 \qquad (5.5)$$

This is exactly what happened when we tried to add (−4) to (+2). The end around carry subtracts off the 2^n and adds a 1 to the result, and we get the correct result. The second example where we tried to add together the two negative numbers is slightly different, and it is shown in Equation 5.6.

$$(-M) + (-N) = \left(2^n - 1 - M \right) + \left(2^n - 1 - N \right) = 2^n + \left[2^n - 1 - (M + N) \right] - 1$$

$$(5.6)$$

This is exactly what happened when we tried to add (−4) to (−2). The end around carry subtracts off the 2^n and adds a 1 to the result, and we get the correct result, which is $(2^n - 1 - (M+N))$, as shown in the square brackets in Equation 5.5.

Arithmetic using 1's complement representation is complicated because we have to account for the end around carry. This arises because there are two representations of zero. Due to the end around carry, sometimes we have to perform two additions to complete one addition. For these reasons, the 1's complement representation is not used for arithmetic and to represent numbers in a computer system.

5.2.4. ARITHMETIC USING 2'S COMPLEMENT REPRESENTATION

To begin, let us look at some examples of adding and subtracting with the digits 2 and 4 when the numbers are written in the 2's complement representation. There are four possible options; these options are shown in Figure 5.7.

```
  2      0010          2      0010
 +4     +0100         -4     +1100
 ──      ────         ──      ────
  6      0110         -2      1110
```

This time, adding the two positive numbers gives us the same result that we got in the previous example, and that is what we expect. This should not be too difficult to understand since we represent the positive numbers in the same way in all three representations of numbers.

```
 -2      1110         -2      1110
 -4     +1100          4     +0100
 ──     ─────         ──     ──────
 -6    1 1010          2    1 0010
```

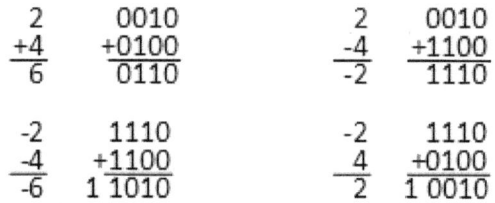

FIGURE 5.7. Adding positive and negative numbers using 2's complement representation.

Adding two negative numbers, we see a major difference. When we add $(-2 \to 1110)$ to $(-4 \to 1100)$, we get a carry out. We throw this carry away. No end around carry here. This gives us the result $(-6 \to 1010)$, which is the correct result. The carry out also occurs when we add (-2) and (4) to each other. Again, we throw away the carry out to get the correct result. In the last addition, there is no end around carry. We throw the carry away so the result of the addition is the expected answer.

Why does the scheme to throw away the carry out work? It works because every time we have a carry out, we have crossed the origin of the number wheel. This time, however, we have only one representation of zero, so we do not have to move the carry to the least significant digit and add it. We can describe when the carry occurs and what is happening if we look at a general addition, as shown in Equation 5.7.

$$M - N = M + \left(\underbrace{2^n - N}_{\substack{\text{Obtaining 1's} \\ \text{complement of N}}} \right) = (M - N) + 2^n \tag{5.7}$$

This is exactly what happened when we tried to add (-4) to (2). The carry just says that we have added 2^n and ignoring the carry subtracts it away, and we get the correct result. The second example where we tried to add together the two negative numbers is slightly different, and it is shown in Equation 5.8.

$$(-M) + (-N) = \left(2^n - M\right) + \left(2^n - N\right) = 2^n + \left[2^n - (M + N)\right] \tag{5.8}$$

This is exactly what happened when we tried to add (-4) to (-2). The carry says that we have added 2^n and ignoring the carry subtracts it away, and we get the correct result $(2^n - (M+N))$, as shown in the square brackets in Equation 5.7.

So the difference between the 1's complement and the 2's complement should now be clear. In the 2's complement, obtaining the negative of a number is a little more complex but the arithmetic is simple. In the 1's complement, obtaining the complement is very simple but the arithmetic is more complex. Since we form the negative of a number only once but may perform arithmetic several times and because the 2's complement has only one representation of the number zero, the 2's complement is the only method used to represent numbers in a computer system.

Review Questions for Section 5.2

Question: Complete the required arithmetic using the sign magnitude representations. Use eight-bit numbers.

1. $32 + 45$;   2) $62 + (-32)$;   3) $(-83) + (-42)$;   2) $(-52) + (47)$

Answer:	32	00100000		62	00111110
	45	00101101		−32	10100000
	77	01001101		30	00011110

	−83	11010011		−52	10110100
	−42	10101010		47	00101111
	−125	11111101		5	10000101

Question: Complete the required arithmetic using the 1's complement representations. Use eight-bit numbers.

1. $32 + 45$;   2) $62 + (-32)$;   3) $(-83) + (-42)$;   2) $(-52) + (47)$

Answer:	32	00100000		62	00111110
	45	00101101		−32	11011111
	77	01001101		30	00011110

	−83	10101100		−52	11001011
	−42	11010101		47	00101111
	−125	10000010		−5	11111010

Question: Complete the required arithmetic using the 2's complement representations. Use eight-bit numbers.

1. 32 + 45;   2) 62 + (−32);   3) (−83) + (−42);   2) (−52) + (47)

Answer:

```
32   00100000          62   00111110
45   00101101        − 32   11100000
─────────────        ──────────────
77   01001101          30   00011110


 −83   10101101        −52   11001100
− 42   11010110         47   00101111
──────────────        ──────────────
−125   10000010         −5   11111011
```

5.2.5. BINARY SUBTRACTION. SUBTRACTION USING N'S COMPLEMENT REPRESENTATION

In the previous section, we saw how we accomplish addition using 2's complement notation to represent numbers. What about subtraction? Subtraction is just as easy as addition. What we have to remember is that when we want to subtract a number, we are changing the sign of that number and then adding it. Examine Equation 5.9.

$$(M)-(N)=(M)+(-N)$$
$$(M)-(-N)=(M)+(N)$$

(5.9)

When we compute the 2's complement of any number, we are simply changing the sign of the number. If the number is a positive number, obtaining its 2's complement makes it a negative number. If the number is a negative number, obtaining its 2's complement makes it a positive number. This is shown in Equation 5.10.

$$(\text{Positive binary number}) \underset{\text{2's complement}}{\Leftrightarrow} (\text{Negative binary number})$$

(5.10)

With this understanding, subtraction is just as easy as addition. There is an additional first step, and that is obtaining the 2's complement of the number to be subtracted; other than that, there is no difference between addition and subtraction in 2's complement representation.

5.3. HALF ADDER AND FULL ADDER USING LOGIC GATES

Now that we know how to do the addition in binary, we have to build a circuit that will do that addition for us. To design and build any logic function, we always begin with a truth table and a logic equation. The truth table for a single-bit adder is given in Figure 5.8. It follows the four rules for binary addition. In the four rules for binary addition, we have two different

X	Y	Sum	Carry
0	0	0	0
0	1	1	0
1	0	1	0
1	1	0	1

$$\text{Sum} = \overline{X} \bullet Y + X \bullet \overline{Y}$$
$$\text{Sum} = X \oplus Y$$
$$\text{Carry} = X \bullet Y$$

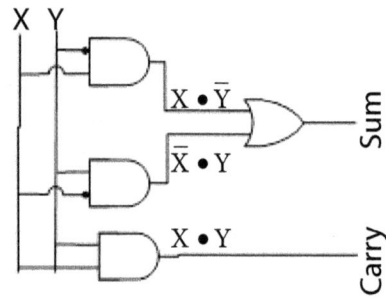

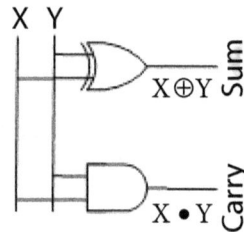

FIGURE 5.8. Truth table, logic Equation and logic diagram for a half adder.

functions. One is for the sum output and the other is for the carry output. The truth table tells us the expected output for the sum output as well as the carry output.

From the truth table, we write out the logic equation for both the sum output and the carry output. This is shown in Equation 5.11.

$$\text{Sum} = \overline{X} \bullet Y + X \bullet \overline{Y} = X \oplus Y$$
$$\text{Carry} = X \bullet Y$$

$$(5.11)$$

Now that we have the logic equation, we can build the logic function from the equation. Notice that we have two different expressions for the sum output so we have drawn two different logic diagrams. Both diagrams perform the same function. All of these are also shown in Figure 5.8. Figure 5.8 shows us what a half adder looks like. It is called a half adder since it adds only the two digits; it does not add the carry that might be present from the previous digit. When we want to add together numbers that are several bits long, we have to include the carry from the previous digit. In these cases, we need a full adder.

The full adder is shown in Figure 5.9. The figure shows the truth table, the logic expression, and the logic diagram. From the truth table, we write

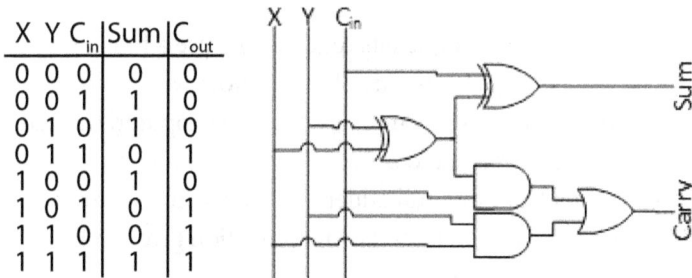

$$\text{Sum} = \overline{X} \bullet \overline{Y} \bullet C_{in} + \overline{X} \bullet Y \bullet \overline{C}_{in} + X \bullet \overline{Y} \bullet \overline{C}_{in} + X \bullet Y \bullet C_{in}$$

$$\text{Sum} = X \oplus Y \oplus C_{in} = \left(X \oplus Y \right) \oplus C_{in}$$

$$\text{Carry} = \overline{X} \bullet Y \bullet C_{in} + X \bullet \overline{Y} \bullet C_{in} + X \bullet Y \bullet \overline{C}_{in} + X \bullet Y \bullet C_{in}$$

$$\text{Carry} = X \oplus Y \bullet C_{in} + X \bullet Y$$

FIGURE 5.9. Truth table, logic Equation and logic diagram for a Full adder.

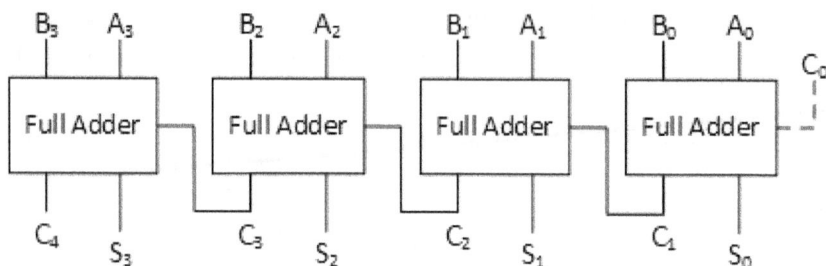

FIGURE 5.10. An Adder block diagram to add together multiple bits.

out the logic equation for both the sum output and the carry output. This is shown in Equation 5.12.

$$\text{Sum} = \overline{X} \bullet \overline{Y} \bullet C_{in} + \overline{X} \bullet Y \bullet \overline{C}_{in} + X \bullet \overline{Y} \bullet \overline{C}_{in} + X \bullet Y \bullet C_{in}$$
$$\text{Sum} = \overline{X} \bullet \left(\overline{Y} \bullet C_{in} + Y \bullet \overline{C}_{in} \right) + X \bullet \left(\overline{Y} \bullet \overline{C}_{in} + Y \bullet C_{in} \right)$$
$$\text{Sum} = \overline{X} \bullet \left(Y \oplus C_{in} \right) + X \bullet \left(Y \oplus C_{in} \right) = X \oplus Y \oplus C_{in}$$
$$\text{Carry} = \overline{X} \bullet Y \bullet C_{in} + X \bullet \overline{Y} \bullet C_{in} + X \bullet Y \bullet \overline{C}_{in} + X \bullet Y \bullet C_{in}$$
$$\text{Carry} = \left(\overline{X} \bullet Y + X \bullet \overline{Y} \right) \bullet C_{in} + X \bullet Y \bullet \left(\overline{C}_{in} + C_{in} \right)$$
$$\text{Carry} = \left(Y \oplus X \right) \bullet C_{in} + X \bullet Y$$

$$(5.12)$$

When we have to add multiple bits together, we use a full adder, as shown in Figure 5.10. In Figure 5.10, we are showing how we would add together two four-bit numbers. In this scheme, C_0 is the carry input to the adder of the least significant digit. This is set to zero since there are no previous digits. The remaining carry bits from one full adder are used as inputs to the adder for the next bit. Each full adder follows the logic equation given in Equation 5.12.

5.4. USING THE ADDER TO DO SUBTRACTION

In Section 5.1.4, we showed that subtraction of binary numbers is performed using 2's complement representation. In this section, we will examine how this is done using the full adder block diagram of Figure 5.10. To obtain the 2's complement of a binary number, we will use the alternative method outlined earlier. In this method, we first complement the binary number and then add 1 to the complemented number. When we are using the adder to perform both the addition and the subtraction, we want to complement the number to be subtracted only when we want to perform subtraction. When we want to perform addition, we do not want to perform the complementation of any number. This condition can be included in the diagram of Figure 5.10, as shown in Figure 5.11.

Figure 5.11 shows a four-bit adder/subtractor that is constructed using the full adder as a building block. In this figure, we show both the inputs B_i and A_i. In addition to this, we also show a new input, which is Add/Subtract. This signal is like a command telling the four-bit adder which operation has to be performed. The command is a zero when addition has to be performed, and it is a one when subtraction has to be performed. Now let us see what happens to the signal A_i when the Add/Subtract command is a zero and when it is a one.

When the command is a zero at that time, the carry input to the least significant bit is a zero and one of the inputs to the XOR gate is also a zero. When one input to the XOR gate is zero, then the output from the XOR gate is equal to the other input, as shown in Equation 5.13. This is exactly what we want when we want to add the two numbers together. We do not want to complement the A_i input and we do not want to add a 1 to the least significant digit.

$$A_i \oplus 0 = A_i \qquad \text{and} \qquad A_i \oplus 1 = \overline{A_i} \qquad\qquad (5.13)$$

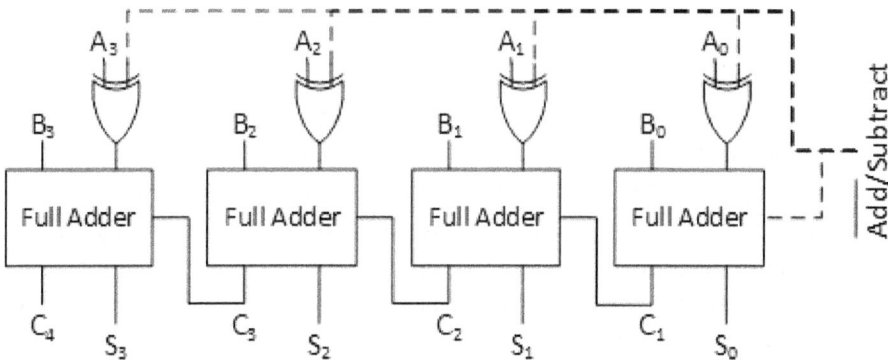

FIGURE 5.11. Using the adder to perform addition as well as subtraction.

When the command is a one, the carry input to the least significant bit is a one and one of the inputs to the XOR gate is also a one. This time, we want to subtract the number A from the number B. To complete the required subtraction, we have to obtain the 2's complement of the number A. This is how it all happens. When one input to the XOR gate is one, then the output from the XOR gate is equal to the complement of the other input, as shown in Equation 5.13. With this, we have achieved the complement portion of determining the 2's complement of the number. Since the carry input to the least significant digit is a 1, this adds a 1 to the least significant digit, which is the same as adding a 1 to the complemented number. Thus, we have successfully obtained the 2's complement of a number that has to be subtracted. Having obtained the 2's complement of the number A, we can proceed to add the two numbers together, in effect subtracting the number A_i from the number B_i.

5.4.1. OVERFLOW IN BINARY ADDITION

We have seen that binary numbers have limited size. In this limited size, we have to represent both positive and negative numbers. Earlier, we saw that we distinguish positive numbers from the negative numbers by examining the most significant bit. When the most significant bit is a 1, then we have a negative number. Examine the addition shown in Equation 5.14.

$$
\begin{array}{r}
0\,1\,1\,0\,1\,0\,1\,0 \\
0\,1\,0\,1\,0\,0\,1\,0 \\
\hline
1\,0\,1\,1\,1\,1\,0\,0
\end{array}
\qquad (5.14)
$$

In Equation 5.14, we have two eight-bit positive numbers that we are adding together. The result of this addition is a negative number. This is not a correct result. What happened? To explain what happened, we will use the number wheel in Figure 5.3 where we have used four-bit numbers. Think of adding numbers as moving around the number wheel in the clockwise direction. So if we tried to add, say, 4 $(0100)_2$ to 5 $(0101)_2$, the result that we would expect is 9 $(1001)_2$, which is what we get, but on the number wheel, the number $(1001)_2$ is listed as -7. This happened because the addition (moving

Carry in the two most significant digits
are different. Overflow has occurred.

```
  5     0 1 0 1      -7     1 0 0 1
 +4    +0 1 0 0      -4    +1 1 0 0
 ───   ───────      ───   ───────
 -7     1 0 0 1      +5     0 1 0 1
```

Carry in the two most significant digits
are same. No overflow has occurred.

```
  5     0 1 0 1      -3     1 1 0 1
 +2    +0 0 1 0      -4    +1 1 0 0
 ───   ───────      ───   ───────
  7     0 1 1 1      -7     1 0 0 1
```

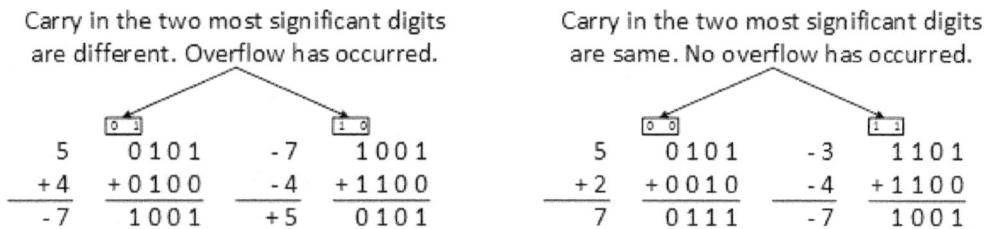

FIGURE 5.12. Overflow occurs when the carry out from the two most significant digits are different from each other.

four places in the clockwise direction from 5) made us cross the number boundary. This can also happen when we are adding two negative numbers. When we are adding negative numbers, we move in the counterclockwise direction. If we tried to add −7 to −4, we would end up with a result of +5.

If we can detect this crossing of the number boundary, then we can correct the result of addition to get the answer that we expect. By the way, we call this an *Overflow*. It is called an overflow because the space available to represent the numbers is not enough and the number is overflowing from the available space. The way we detect the overflow is shown in Figure 5.12. We see that an overflow occurs when the carry in to the most significant digit is different from the carry out from the most significant digit. This is shown in the two additions on the left side of Figure 5.12. On the right side of Figure 5.12, we see that no overflow occurs because the carry in to the most significant digit is equal to the carry out of the most significant digit. If we take this idea to our adder, adding an XOR gate that has the two carries as its input will allow us to detect when an overflow has occurred.

5.5.   COMMERCIAL ARITHMETIC UNIT

Adders used in arithmetic circuits follow the general design we have seen in Figure 5.11 with one major difference. To understand why we need this major difference, we must first see what drawback in the adder circuits these commercial IC's alleviate. For this, consider the time it takes the carry to go through a full adder circuit. This is shown in Figure 5.13. The circuit

X Y C_in

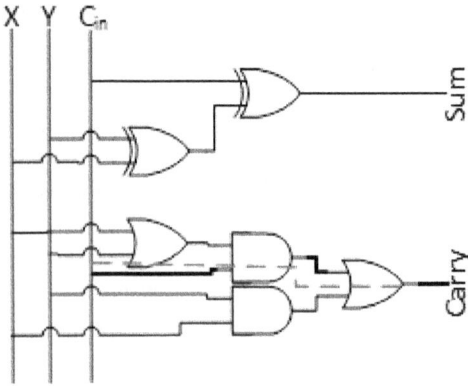

Sum

Carry

FIGURE 5.13. Signal path that the Carry input takes through a full adder.

diagram for the full adder is slightly different from the one shown earlier since we wanted to separate the path that the sum signal takes and the path that the carry signal takes. We see that the longest path that carry signal takes is a path that goes through three gates. Every gate that the carry signal has to go through introduces a delay in the carry signal. The typical delay through any one gate is about 10 nSec. With this delay, we can complete the addition of one bit with a delay that would be 30 nSec. Only after this carry has been computed can we use it in the addition of the next bit. The addition of the next bit begins 30 nSec after the start of the addition of the first bit. We say that the carry has to *ripple* through all the bits one bit at a time. In a typical computer, there are thirty-two bits in any number, so it is possible that the carry has to ripple through all these thirty-two bits to complete the addition. This would take 32 * 30 = 960 nSec. (It is true that this will not be the case every time, but you have to account for the worst case. Since you do not know beforehand when this will occur, you have to assume that if and when it occurs there will be enough time to complete the addition.) If your computer spent 960 nSec for every addition, the speed of the computer could not be more than 1 MHz. That is very slow. So the engineers have used some tricks to speed up the addition of a long number.

5.5.1. CARRY LOOK AHEAD LOGIC

Let us consider the case of a four-bit adder. We will express the carry out of each stage as a function of the two input bits and the carry in from the previous bit. The idea of *carry look ahead* is to express each carry bit in terms of the inputs prior to the current bit (these are X_i, X_{i-1}, X_{i-2}, ..., X_0, are Y_i, Y_{i-1}, Y_{i-2}, ..., Y_0) and the initial carry in C_0. The carry out logic equation is a very complicated logic function, but as we have seen earlier, we can express any logic function as a two-level AND-OR-Invert expression. Representing the carry out logic function this way as a two-level logic, the carry should never take more than two gate delays to complete.

We begin the design of carry look ahead adders by defining two additional functions that will help us. They are the carry generate, identified as G_i, and carry propagate, identified as P_i. They are defined as shown in Equation 5.15.

$$G_i = X_i \bullet Y_i \qquad P_i = X_i \oplus Y_i \qquad (5.15)$$

From our previous study, we know that whenever X_i and Y_i are both 1, then there will always be a carry out from that bit to the next bit. That is what the generate function does. It checks all the X_i and the Y_i simultaneously and produces a carry out wherever it is required. The other function is a carry propagate; that is, it will transfer the same state of the carry from bit i to bit i+1. This happens when only one of the X_i or Y_i is a 1. So, when the XOR operation gives us a logic High output, then the carry out will be the same as the carry in. The next step that we will take is to define both the sum output and the carry output in terms of the carry generate and the carry propagate functions, as shown in Equation 5.16.

$$\begin{aligned}
S_i &= X_i \oplus Y_i \oplus C_i = P_i \oplus C_i \\
C_{i+1} &= X_i \bullet Y_i + X_i \bullet C_i + Y_i \bullet C_i \\
&= X_i \bullet Y_i + C_i \bullet (X_i + Y_i) \\
&= X_i \bullet Y_i + C_i \bullet (X_i \oplus Y_i) \\
&= G_i + C_i \bullet P_i
\end{aligned} \qquad (5.16)$$

So when the carry out from stage i is a 1, then it is either internally generated (G_i=1) or the carry into the stage (C_i) was a 1 and the carry is propagated (P_i=1) out to the next stage. When we express the carry out this way, we can write the expression for the carry for each stage as shown in Equation 5.17.

$$\begin{aligned}
C_1 &= G_0 + P_0 C_0 \\
C_2 &= G_1 + P_1 C_1 = G_1 + P_1(G_0 + P_0 C_0) = G_1 + P_1 G_0 + P_1 P_0 C_0 \\
C_3 &= G_2 + P_2 C_2 = G_2 + P_2 G_1 + P_2 P_1 G_0 + P_2 P_1 P_0 C_0 \\
C_4 &= G_3 + P_3 C_3 = G_3 + P_3 G_2 + P_3 P_2 G_1 + P_3 P_2 P_1 G_0 + P_3 P_2 P_1 P_0 C_0
\end{aligned} \qquad (5.17)$$

In Equation 5.15, C_1 is the carry out from the least significant digit and so on for the other carries. Equation 5.15 shows us that in principle, we can get any carry out using a SOP circuit that is of only two gate delays. When you examine the equations carefully, you will see that the carry out C_4 has five terms that have to be ORed together. One of these terms is the output of an AND gate that has five terms as input. In general, the carry out from stage i will have i+1 terms ORed together and one of the terms will be from a gate that has i+1 inputs to it. Due to this limitation, we find plenty of examples of four-stage look ahead circuits. Carry look ahead circuits for more than four bits are very rare. To build an eight-stage look ahead circuit is very difficult because of the scarcity of a nine input AND gate and a nine-input OR gate.

5.5.2. BUILDING A CARRY LOOK AHEAD ADDER

The basic idea of a carry look ahead adder is shown in Figure 5.14. This represents the full adder. To the full adder we need to add the carry look ahead circuits. The carry look ahead circuits for a four-bit adder are shown in Figure 5.15. From Figure 5.15, we can now determine the time it takes to complete the addition for four-bit numbers. In Figure 5.14, we have seen that it takes one gate delay to produce the propagate and the generate signal and a two gate delay to produce the sum signal. Signals C_1 C_2 C_3 and C_4 will be ready two gate delays after the propagate and the generate signals are produced, so all the carry signals are ready after three gate delays. The sum signals are ready one gate delay after the carry signals are ready, so S_1 S_2 and S_3 will be ready one gate

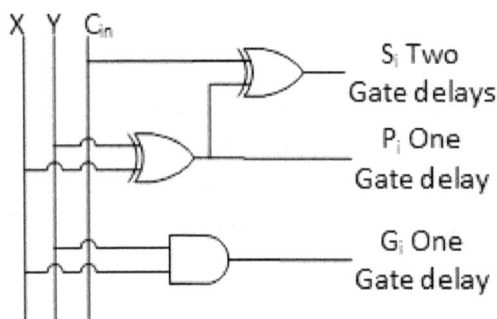

FIGURE 5.14. Carry Look ahead full adder with propagate and generate.

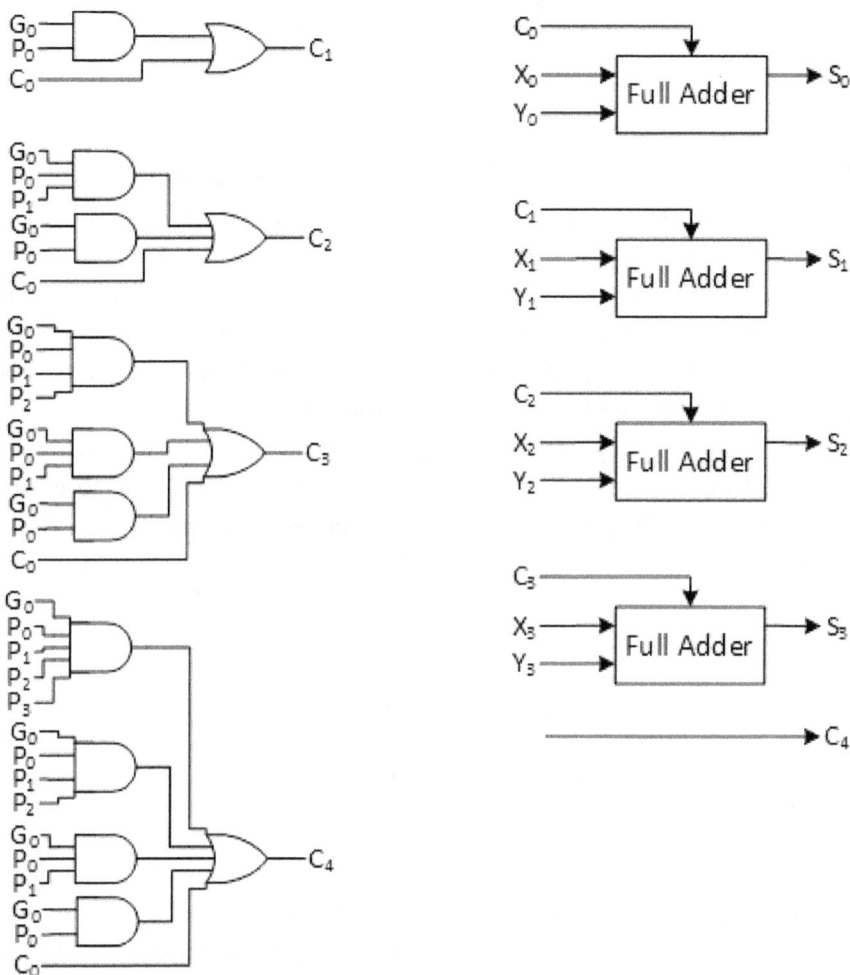

FIGURE 5.15. A 4-bit Carry look ahead circuits and a 4 bit adder.

delay after the carry signals, hence a total of four gate delays. This is definitely much faster than the ripple carry through the four-bit adder that would have taken up to twelve gate delays. The price that we pay here is added complexity in the circuit and the need for gates with a larger number of inputs. As we increase the number of inputs to a gate, the speed of the gate degrades, so a five-input logic gate is not as fast as a two-input logic gate.

5.6. BCD NUMBERS AND BCD ADDITION

BCD or *binary coded decimal* representation uses four binary digits to represent a decimal digit. When you use four binary digits, you have available sixteen different representations. If you are only using ten representations, you are wasting space. But the notion that you can group binary digits in groups of four and know what decimal digit that group represents is intriguing. Also, in the early days of digital hardware, this presented an easy way to convert the result of any computation into a digital display. Today, we have had enough advances in digital hardware and digital logic that the early advantage of being able to easily display decimal digits has all but disappeared. Here, we present the BCD representation and the hardware to do arithmetic only as an academic exercise.

5.6.1. REPRESENTING NUMBERS USING THE BCD SCHEME

In the BCD representation, we use only the decimal digits. These digits are the numbers from 0 to 9. These ten digits are represented using four binary digits. The four binary digits 0000_2 correspond to the decimal digit 0 through the four binary digits 1001_2, which corresponds to the decimal digit 9. The remaining six binary combinations 1010_2 to 1111_2 are not used and hence treated as don't care. Addition using BCD representation proceeds just as we perform arithmetic in our everyday life, where we use the decimal number system. This time, however, we have to be careful when there is a carryover from one digit to another. We have to first detect the condition when a carryover will occur and then have the circuitry present that will accomplish the carryover.

As an example, consider the addition of two BCD digits as shown in Figure 5.16.

In Figure 5.16, the addition of 3 to 4 gives us a correct result. This shows us that the addition is correct when there is no carryover to the next digit. Now consider the addition of 5 and 8. The sum 1101_2 is indeed 13, but it is not a BCD digit, so it

$$3 = 0\,0\,1\,1$$
$$\underline{4 = 0\,1\,0\,0}$$
$$7 = 0\,1\,1\,1$$

Add 3 and 4 together

$$5 = 0\,1\,0\,1$$
$$\underline{8 = 1\,0\,0\,0}$$
$$13 = 1\,1\,0\,1$$

Add 5 and 8 together

FIGURE 5.16. Performing addition in BCD.

will not be correctly represented. The sum 13 should be represented as two BCD digits, which will be $(0001\ 0011)_2$. Fortunately, we can correct this in a very simple manner. The procedure that we use to correct is to add 0110_2, which is a 6, to the result whenever the result exceeds 1001_2, which is a 9. Let us see how this works; look at Figure 5.17.

The result of both the additions in Figure 5.17 exceed the digit 9; this requires that we add 0110_2 or 6 to the result. This is to bypass the six don't care terms we have when we represent the decimal numbers using BCD representation. With the addition of the 6, we get the result in two BCD digits. That is what we expected, and that is what we got.

5.6.2. THE BCD ADDER

Figure 5.18 shows us the design of a BCD adder. In the figure, the top row of four full adders performs the regular binary addition. The second row of two adders provides us the capability of adding $(0110)_2$ when it is required, and at all other times we add $(0000)_2$. Remember we have to add $(1001)_2$ when the sum of binary addition exceeds $(1001)_2$.

$$5 = \quad 0\,1\,0\,1$$
$$\underline{8 = \quad 1\,0\,0\,0}$$
$$13 = \quad 1\,1\,0\,1 \quad \text{13 in Binary}$$
$$\underline{6 = \quad 0\,1\,1\,0}$$
$$13 = 1\,0\,0\,1\,1 \quad \text{13 in BCD}$$

Add 5 and 8 together

$$9 = \quad 1\,0\,0\,1$$
$$\underline{8 = \quad 1\,0\,0\,0}$$
$$17 = {}^1 0\,0\,0\,1 \quad \text{17 in Binary}$$
$$\underline{6 = \quad 0\,1\,1\,0}$$
$$17 = 1\,0\,1\,1\,1 \quad \text{17 in BCD}$$

Add 9 and 8 together

FIGURE 5.17. Correcting for Carry in BCD arithmetic.

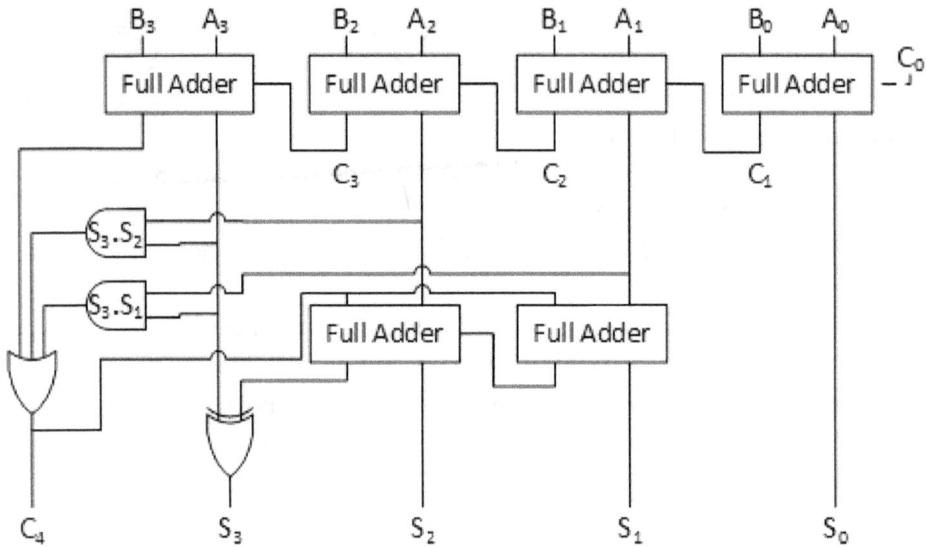

FIGURE 5.18. A BCD Adder.

The rest of the circuit is used to detect when the sum of the binary addition exceeds $(1001)_2$. We need to add $(0110)_2$ when there is a carry out of the $A_3 B_3$ full adder (binary sum exceeds 15), when S_3 and S_2 are both logic High (binary sum is $(11xx)_2$), or when both S_3 and S_1 are logic high (binary sum is $(1X1x)_2$). These conditions are detected by the two AND gates and combined together by the OR gate. When the output of the OR gate is a 1, then we have a carryover to the next digit, so C_4 is a 1. This 1 is used as input to the two AND gates, which adds $(0110)_2$ to the result of the binary addition. This adder requires substantially more hardware and there is greater time delay. Since we have available substantially faster and cheaper adders, it is no surprise that the idea of using the BCD number system and BCD adders was dropped very quickly.

5.7. CHAPTER PROBLEMS

5.7.1.   Represent the following numbers in the indicated number system. Assume that the numbers have to be represented in an eight-bit number.

5;   22;   -69;   88;   -144;   123.

5.7.1.1. Represent the numbers in sign magnitude representation.

5.7.1.2. Represent the numbers in 1's complement representation.

5.7.1.3. Represent the numbers in 2's complement representation.

5.7.2. Perform the indicated arithmetic assuming that the numbers are represented in sign magnitude notation.

0 1 1 0 1 1 0 0 0	1 1 1 0 0 0 1 1 1	0 0 0 1 1 1 0 1
1 1 1 1 0 0 0 1 0	1 1 0 1 1 0 1 0 1	1 0 0 1 1 1 0 0

5.7.3. Perform the indicated arithmetic assuming that the numbers are represented in 1's complement notation.

0 1 1 0 1 1 0 0 0	1 1 1 0 0 0 1 1 1	0 0 0 1 1 1 1 0
1 1 1 1 0 0 0 1 0	1 1 0 1 1 0 1 0 1	1 0 0 1 1 1 0 1

5.7.4. Perform the indicated arithmetic assuming that the numbers are represented in 2's complement notation.

0 1 1 0 1 1 0 0 0	1 1 1 0 0 0 1 1 1	0 0 0 1 1 1 1 0
1 1 1 1 0 0 0 1 0	1 1 0 1 1 0 1 0 1	1 0 0 1 1 1 0 1

5.7.5. We want to build a logic circuit that will convert a 1's complement representation to 2's complement representation for four-bit numbers. Build a truth table and then write the logic equation in the SOP representation and in the POS representation.

5.7.6. We want to build a logic circuit that will convert a 2's complement representation to 1's complement representation for four-bit numbers. Build a truth table and then write the logic equation in the SOP representation and in the POS representation.

5.7.7. We want to build a logic circuit that will convert a 1's complement representation to sign magnitude representation for four-bit numbers. Build a truth table and then write the logic equation in the SOP representation and in the POS representation.

5.7.8. We want to build a logic circuit that will convert a 2's complement representation to sign magnitude representation for four-bit numbers. Build a truth table and then write the logic equation in the SOP representation and in the POS representation.

5.7.9. We want to build a logic circuit that will convert a sign magnitude representation to 2's complement representation for four-bit numbers. Build a truth table and then write the logic equation in the SOP representation and in the POS representation.

5.7.10. Using the subtraction tables given in Figure 5.4, develop a truth table for a full subtractor. This should include both the borrow logic as well as subtract logic.

5.7.11. Earlier, we mentioned the difficulties with a sign magnitude adder. Here, we will assume that we are always adding numbers that are the same sign so the difficulties are minimum. Design an adder for a four-bit sign magnitude adder. Remember this has to be a full adder.

5.7.12. Most adders are used to perform logic functions as well as adding operations. Here, we will modify the inputs to the four-bit adder so that we can perform other logic functions as well as addition. To select the operation to be performed, we will use a three-bit input code. We will be able to perform eight different functions, as described in the table below.

F_2	F_1	F_0	Operation to be performed
0	0	0	$S_3 S_2 S_1 S_0 = 0\,0\,0\,0$
0	0	1	$S_3 S_2 S_1 S_0 = (B - A)$
0	1	0	$S_3 S_2 S_1 S_0 = (A - B)$
0	1	1	$S_3 S_2 S_1 S_0 = (B + A)$
1	0	0	$S_3 S_2 S_1 S_0 = (B \text{ AND } A)$
1	0	1	$S_3 S_2 S_1 S_0 = (B \text{ OR } A)$
1	1	0	$S_3 S_2 S_1 S_0 = (B \text{ XOR } A)$
1	1	1	$S_3 S_2 S_1 S_0 = 1\,1\,1\,1$

5.7.13. Perform the indicated arithmetic assuming that the numbers are represented in BCD notation.

```
0 1 1 0 1 0 0 1 0      1 1 1 0 1 0 1 1 0      0 0 0 1 1 1 1 1
1 0 0 1 0 0 0 1 0      1 1 0 1 0 0 1 0 1      1 0 0 1 0 1 0 0
```

6. COMMON COMBINATIONAL LOGIC FUNCTIONS

6.0. INTRODUCTION

In the earlier chapters, we examined how we can use simple logic gates to build a logic function. At that time, we also studied how we can reduce the cost of the logic function. The fundamental idea in reducing the cost of the logic function was to reduce the number of connections on the printed circuit board. Another way to reduce the number of connections on the outside is to use a higher level of integration inside the integrated circuit. When we increase the level of integration inside the IC, we fix the function that the IC can perform. This is good if the function to be performed is a common function. There are only a few very common functions. We begin this chapter by looking at these functions.

The predefined functions are few, but the advantages gained from using an entire logic circuit in a single IC are many. Following up on these advantages, the engineers have come up with a method to design ICs so that each can be custom designed to the need of the user. Toward the end of the chapter, we will study how programmable logic can be used to custom design the logic function that you need.

6.1. FREQUENTLY USED LOGIC FUNCTIONS

Earlier, we decided that we were going to build all logic functions so that we get the output after only two gate delays. All the functions we built using the two-gate delay limit gave us either the SOP or the POS logic functions. These logic functions are very regular; for example, the SOP logic function always has several AND gates, which are the first level of gates, and the output of the AND gates are combined together by one OR gate, which is the second level of the function. Similarly, when the logic function is written in the POS form, we have several OR gates, which are the first level of gates, and the output of the OR gates are combined together by one AND gate, which is the second level of the function. Writing the logic function in either of the two forms is the same as far as the gate delay and the form of the function are concerned. For this reason, the engineers have built several functions that follow this exact format.

6.1.1. MULTIPLEXERS

The first such function that we will examine is the multiplexer. It is also known as the data selector, as it can be used to choose one input out of many to be connected to the output depending on the control signal. To see this, examine Figure 6.1. In Figure 6.1, we have four different data inputs and one data output. Along with the inputs and outputs, we also have a switch that is controlled by the data selector inputs. The data selector inputs position the switch such that one and only one input is directly connected to the output. In the figure, we see that when the data selector inputs are $(0\,0)_2$, then the data input D_0 will be connected to the output. Similarly, we can see that for selector inputs of $(1\,0)_2$, the data input D_2 will be directly connected

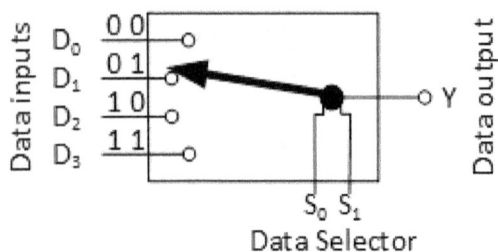

FIGURE 6.1. Demonstrating the operation of a multiplexer.

to the output. We extend this same concept to describe a logic function, as shown in Figure 6.2. Figure 6.2 shows us the truth table and the logic diagram of a logic multiplexer. Here, we use the data selector inputs S_0 and S_1 as inputs to AND gates. These AND gates function as the switches that choose the input to be connected to the output. So when the data selectors $S_1 S_0 = (0\ 1)_2$, then the AND gate labeled "1" is selected and the output from the multiplexer is the D_1 input. The truth table shows

D	S_1	S_0	F
D_0	0	0	D_0
D_1	0	1	D_1
D_2	1	0	D_2
D_3	1	1	D_3

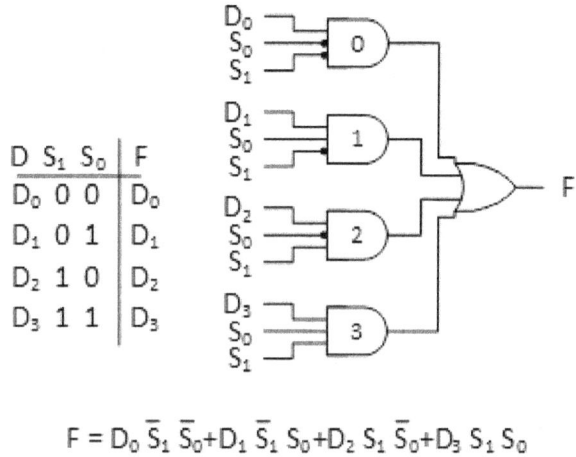

$$F = D_0\ \overline{S_1}\ \overline{S_0} + D_1\ \overline{S_1}\ S_0 + D_2\ S_1\ \overline{S_0} + D_3\ S_1\ S_0$$

FIGURE 6.2. Demonstrating the operation of a multiplexer.

us when each of the AND gates are selected and what the corresponding output will be when the AND gate is selected.

Multiplexers come in various different sizes. They are identified by the number of inputs and the number of outputs, so the multiplexer in Figure 6.2 will be classified as a 4-to-1 multiplexer. The number of inputs also tells us how many AND gates we need to build the multiplexer. We need one AND gate for each input that is present. The number of outputs tells us how many OR gates we need. We need one OR gate for each unique output from the multiplexer.

Commercially, we can purchase several different multiplexers. For example, the 74HC157 is a quad two-input multiplexer. This means that on the 74HC157, there are four different multiplexers; each of them has two inputs and one output. The 74HC151 is an 8-to-1 multiplexer. It has eight inputs and one output. The 74HC150 is a 16-to-1 multiplexer. Examine Figure 6.3 to look at some details of a commercial multiplexer like the 74HC151.

This multiplexer has eight data inputs, so it needs three select inputs. There are also two outputs. These two outputs are not different, unique outputs. One is just the complement of the other. There is one additional input. This is the enable input. The enable input is like a "gate" to the IC. If the enable input is not asserted, then the IC is not selected. This is like

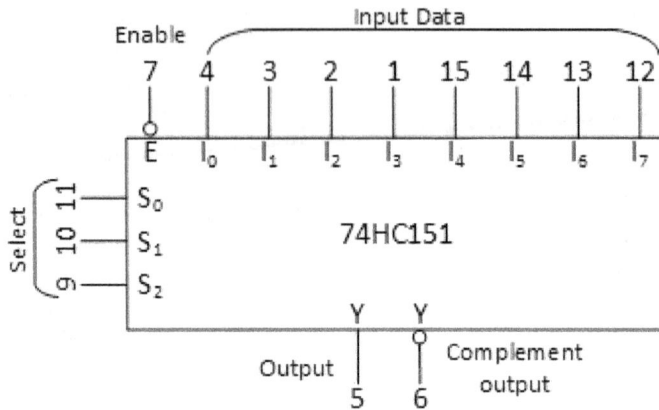

FIGURE 6.3. Pin diagram of 74HC151 8– to –1 multiplexer.

saying that the gate is closed and no signals can enter. This means no inputs and no data select signals can go inside the IC. Therefore, to use the IC, we have to first enable the IC. This time, since there is a "bubble" at the input of the enable signal, we must have a logic Low signal present at the enable input pin.

Inside this multiplexer we will find eight AND gates. Each AND gate gets as its input the three data select inputs, one of the data inputs, and the enable input. The output of all the eight AND gates are connected to the input of an eight-input OR gate. This is the output from the multiplexer.

Question: In Figure 6.4, if the switch inputs are $(1\ 1)_2$, which data input will be connected to the output?

Answer: With the switch input set to $(1\ 1)_2$, the input D_3 will be connected to the output.

Question: Which input will be connected to the output if the switch inputs are $S_1S_0 \rightarrow (0\ 0)_2$ at time t_0; change to $S_1S_0 \rightarrow (1\ 0)_2$ at time t_1; change to $S_1S_0 \rightarrow (0\ 1)_2$ at time t_2; and change to $S_1S_0 \rightarrow (1\ 1)_2$ at time t_3 in Figure 6.4?

Answer: With the switch input set to $(0\ 0)_2$, the input D_0 will be connected to the output. When the switch inputs change to $S_1S_0 \rightarrow (1\ 0)_2$ at time t_1, the input D_2 will be connected to the output. When the switch inputs change to $S_1S_0 \rightarrow (0\ 1)_2$ at time t_2, the input D_1 will be connected

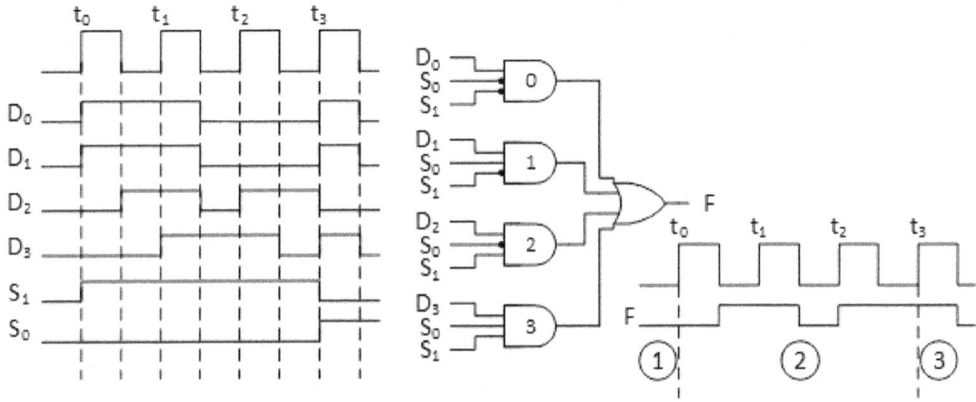

FIGURE 6.4. Demonstrating the operation of a multiplexer.

to the output. When the switch inputs change to $S_1 S_0 \rightarrow (1\ 1)_2$ at time t_3, the input D_3 will be connected to the output.

6.1.1.1. USING THE MULTIPLEXER

The simplest use of the multiplexer is to use it as a data selector. This use of the multiplexer is shown in Figure 6.4. On the left hand side of Figure 6.4, we see a timing pulse signal with times t_0, t_1, etc. shown to identify the different times. Below the timing signal, we see the four different input signals: D_0, D_1, D_2, and D_3. Below the input signals, we see the select signals. All these signals are linked with the timing signals on the top. Assuming that these are input signals, we wish to draw the diagram of what the output signal will look like. This is how we analyze the multiplexer.

Before time marker t_0, we see that the select inputs $S_1 S_0$ are $(0\ 0)_2$. During this time, the input D_0 is connected to the output and the output is the same as the input D_0. This is marked by marker ① on the output side. At time marker t_0, the select input S_1 changes to 1. Now the two select inputs $S_1 S_0$ are $(1\ 0)_2$. This will connect the input D_2 to the output so the output is the same as the input D_2. This is marked by marker ② on the output side. At time marker t_3, select input S_0 changes to 1 and select input S_1 changes to 0. Now the two select inputs $S_1 S_0$ are $(0\ 1)_2$. This will connect the input D_1 to the output so the output is the same as the input D_1. This is marked by marker ③ on the output side.

6.1.1.2.   USING A MULTIPLEXER TO BUILD A COMBINATIONAL LOGIC FUNCTION

Another important function that a multiplexer provides is to build a combinational logic function. To see how this works, we need to understand Shannon's Expansion Theorem.

Shannon's Expansion Theorem: Shannon's Expansion Theorem allows us to separate any logic function into two different parts. In one part of the function, one variable is always in its true form. In the other part of the function, the same variable is always in its complement form. Since the variable in one part always appears in its true form, we can factor that variable out of this part of the function. Similarly, since the variable in the other part always appears in its complement form, we can factor that variable out of this part of the function. This is shown in Equation 6.1 with the variable x_1 being the variable that is factored out.

$$f(x_1 x_2 \cdots x_n) = x_1 f'(1 x_2 \cdots x_n) + \overline{x}_1 f''(0 x_2 \cdots x_n) \qquad (6.1)$$

Equation 6.1 shows a function f of several variables. We wish to expand this function about variable x_1. The result of the expansion is shown on the right hand side. The function f' is the part of the function that has variable x_1 in its true form, so we have factored it out of the function f'. The remaining part of the function f is the function f'' in which the variable $\overline{x}_1$ appears in its complement form, so we have factored it out of the function f''. Of course, there is nothing stopping us from using the Expansion Theorem once again on functions f' and f''. Expanding it a second time, we cannot expand it over the same variable, as that variable has been factored out, but we can expand the two functions about another variable. The expansion would have to be over a different variable. Equation 6.2 shows us an example of using Shannon's Expansion Theorem.

$$f = \overline{x} \bullet \overline{y} \bullet \overline{z} + \overline{x} \bullet \overline{y} \bullet z + \overline{x} \bullet y \bullet \overline{z} + x \bullet \overline{y} \bullet z + x \bullet y \bullet z$$
$$f = x \bullet \left(\overline{y} \bullet z + y \bullet z \right) + \overline{x} \bullet \left(\overline{y} \bullet \overline{z} + \overline{y} \bullet z + y \bullet \overline{z} \right) \quad (6.2)$$

Notice in Equation 6.2 we have expanded the function f about the variable x so the function f' and the function f'' do not have the variable x present in them; the variable x is factored out.

Question: Use Shannon's Expansion Theorem to expand the function given in Equation 6.2 about the variable Y.

Answer: The expansion about variable Y is $f = y \bullet \left(\overline{x} \bullet \overline{z} + x \bullet z \right) + \overline{y} \bullet \left(\overline{x} \bullet \overline{z} + \overline{x} \bullet z + x \bullet z \right)$. This can be further simplified to $f = y \bullet \left(\overline{x} \bullet \overline{z} + x \bullet z \right) + \overline{y} \bullet \left(\overline{x} + z \right)$.

Question: Use Shannon's Expansion Theorem to expand the function given in Equation 6.2 about the variable Z.

Answer: The expansion about variable Z is $f = z \bullet \left(\overline{x} \bullet \overline{y} + x \bullet \overline{y} + x \bullet y \right) + \overline{z} \bullet \left(\overline{x} \bullet y + \overline{x} \bullet \overline{y} \right)$. This can be further simplified to $f = z \bullet \left(\overline{y} + x \right) + \overline{z} \bullet \left(\overline{x} \right)$.

6.1.1.3. USING A MULTIPLEXER TO BUILD A LOGIC FUNCTION

To use the multiplexer to build a logic function, examine Figure 6.5. In Figure 6.5, we want to build a logic function $f = \overline{x} \bullet y + x \bullet \overline{y}$ using the multiplexer. To do this, we first choose one variable that will be used as a switch variable. In the example, we have chosen the variable x to represent the switch variable. Next, we expand the function about the switch variable; in this example, we have used the variable x for this purpose. This will separate the function f into two sub-functions. One of the sub-functions has the switch variable represented in its true form, while the other sub-function has the switch variable represented in its complemented form. Therefore, to build the function, we will connect the variable x to the switch input. To the

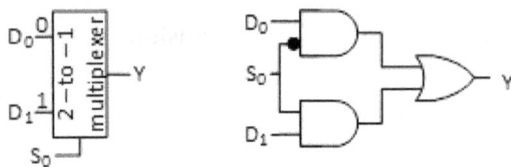

$$f = \overline{x} \bullet y + x \bullet \overline{y}$$
$$f = x \bullet \left(\overline{y}\right) + \overline{x} \bullet \left(y\right)$$

$$S_0 = x$$
$$D_0 = y \qquad D_1 = \overline{y}$$

FIGURE 6.5. Using the Multiplexer to build a logic function.

input D_0 we will connect the function from Shannon's Expansion Theorem that is associated with the variable in its complemented form. This is because when the switch input is in its complemented form D_0 is connected to the output. To the input D_1 we will connect the function from Shannon's Expansion Theorem that is associated with the variable in its true form. This is because when the switch input is in its true form D_1 is connected to the output. All this is shown in Figure 6.5.

What if we have a function with more variables and a multiplexer with several switch inputs, say, like the multiplexer in Figure 6.4? There, we have a multiplexer with two switch inputs. To use this multiplexer, we will choose two of the variables from the function to be the switch inputs. Say we want to use a 4-to-1 multiplexer to build the logic function given in Equation 6.3. This time, since we have two switch inputs, we will use Shannon's Expansion Theorem twice. First with the variable x. This will give us two smaller functions. Next, for each of the two smaller-functions we will use Shannon's Expansion Theorem with variable y. This gives us four different functions.

$$f = \overline{w} \bullet x \bullet y + \overline{w} \bullet \overline{x} \bullet y + \overline{w} \bullet x \bullet \overline{y} + w \bullet \overline{x} \bullet \overline{y} + w \bullet x \bullet \overline{y}$$

$$f = x \bullet \left(\overline{w} \bullet y + \overline{w} \bullet \overline{y} + w \bullet \overline{y}\right) + \overline{x} \bullet \left(\overline{w} \bullet y + w \bullet \overline{y}\right)$$

$$f = x \bullet \left(y \bullet \underbrace{\left(\overline{w}\right)}_{D_3} + \overline{y} \bullet \underbrace{\left(\overline{w} + w\right)}_{D_1} \right) + \overline{x} \bullet \left(y \bullet \underbrace{\left(\overline{w}\right)}_{D_2} + \overline{y} \bullet \underbrace{\left(w\right)}_{D_0} \right) \qquad (6.3)$$

The function identified as D_0 is connected to the input D_0 on the multiplexer since both the switch variables have to be in their complemented form to select this function. In the same way, the function identified as D_1 has its switch variables as $x\, y \to (0\ 1)_2$.

Question: We want to implement the following function using a multiplexer. Choose variables X and Z as switch variables.

$$f = x \bullet \overline{y} \bullet z + \overline{x} \bullet \overline{y} \bullet z + \overline{x} \bullet y \bullet \overline{z} + x \bullet y \bullet z + \overline{x} \bullet \overline{y} \bullet \overline{z}$$

Answer: We first expand the function about the variable X and then about the variable Z. After the expansion is complete, we can simplify the remaining function, as shown here.

$$f = x \bullet \left(\overline{y} \bullet z + y \bullet z \right) + \overline{x} \bullet \left(\overline{y} \bullet z + y \bullet \overline{z} + \overline{y} \bullet \overline{z} \right)$$

$$f = x \bullet \left(z \bullet \left(y + \overline{y} \right) + \overline{z} \bullet \left(0 \right) \right) + \overline{x} \bullet \left(z \bullet \left(\overline{y} \right) + \overline{z} \bullet \left(y + \overline{y} \right) \right)$$

$$f = x \bullet \left(z \bullet \left(1 \right) + \overline{z} \bullet \left(0 \right) \right) + \overline{x} \bullet \left(z \bullet \left(\overline{y} \right) + \overline{z} \bullet \left(1 \right) \right)$$

Now we have to match each of the functions to the proper input. To do this, we first connect the variable x to the switch input S_0 and the variable y to the switch input S_1. When the switch inputs are $(0\ 0)_2$, the input D_0 is selected by the multiplexer. Therefore, we connect to the input D_0 the function from Shannon's expansion that is selected when both x and y are in their complemented form. This is shown in Equation 6.3. In the same manner, we can match up the switch conditions to the Shannon expansion variables to build the logic function using a multiplexer.

6.1.2. DE-MULTIPLEXERS

This is the opposite of a multiplexer. This time, we have one input and the data selector. The data selector switch now steers the input to one of the many different outputs; this operation is shown in Figure 6.6. Since there is only one input and several outputs, the logic equation for each output will have the input term as part of it. This is shown in Figure 6.6. To build the de-multiplexer, we need as many AND gates as there are outputs. When a particular AND gate is selected by the switch inputs, the output from that AND gate is the same as the input to the de-multiplexer.

The integrated circuit de-multiplexers come in several differ-ent configurations of input-output

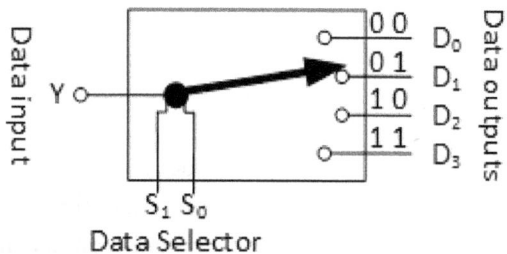

FIGURE 6.6. Demonstrating the operation of a de-multiplexer.

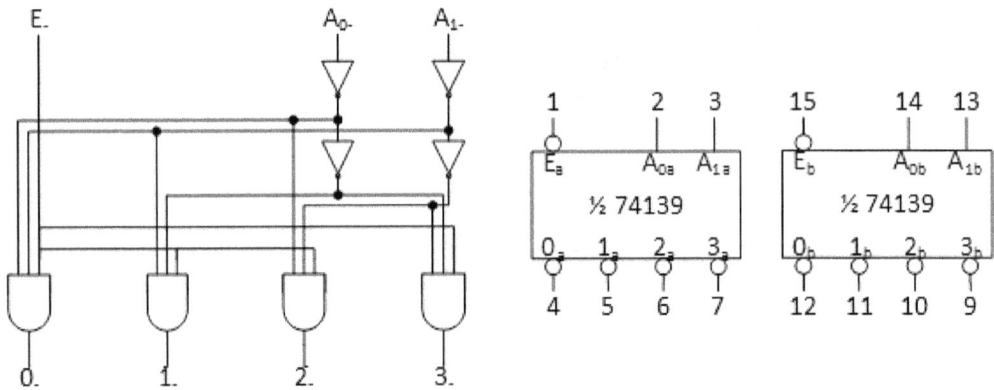

FIGURE 6.7. Pin diagram and Logic diagram for a de-multiplexer.

combinations. One of these de-multiplexers is shown in Figure 6.7. There, we see the 74139 de-multiplexer; this is a dual 4-to-1 de-multiplexer. From the figure, we see that the two halves of the multiplexer are identical to each other. The logic that performs the de-multiplexing function is also shown in Figure 6.7. To use the de-multiplexer, the input is provided on the E input for each half of the de-multiplexer. Then, using the switch inputs A_0 and A_1, the input is steered to the appropriate output. From Figure 6.7, we see that the input E is the input to all the four AND gates. The other two inputs to the AND gates are the switch inputs. When the switch inputs select one particular AND gate, then the input E will be the output from that same AND gate while all the other AND gates will put out only a logic 0 since those AND gates were not selected by the switch inputs. This way, we steer the input to any one of the outputs.

6.1.3. DECODERS

Decoding is a process of converting information that is in one format to the same information in a different format. An example of a decoder would be an IC that will take the binary input and convert it to BCD format. There are many such examples of code conversions, and all these conversions can be completed with the use of a decoder that is designed for the specific purpose. Let us examine one such decoder. This decoder converts a binary number into its octal equivalent. The truth table for such a decoder is shown in Figure 6.8. Examine the inputs and the outputs in the truth table. For each row of a unique input combination, we have eight different outputs. Of these eight outputs, seven of them are identical and the eighth output is different. We say the decoder has decoded the input and the decoded value is that of the unique output. Notice that there is only one output that is decoded for one combination of the input. In the example, the decoded output is indicated by a logic level that is High while all the other outputs have a Low logic level. It is also possible to have the unique logic level that is Low and all the other logic levels are High.

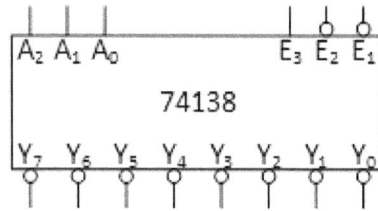

Truth Table for Binary to Octal Decoder.

Input			Output							
A_2	A_1	A_0	Y_7	Y_6	Y_5	Y_4	Y_3	Y_2	Y_1	Y_0
0	0	0	0	0	0	0	0	0	0	1
0	0	1	0	0	0	0	0	0	1	0
0	1	0	0	0	0	0	0	1	0	0
0	1	1	0	0	0	0	1	0	0	0
1	0	0	0	0	0	1	0	0	0	0
1	0	1	0	0	1	0	0	0	0	0
1	1	0	0	1	0	0	0	0	0	0
1	1	1	1	0	0	0	0	0	0	0

FIGURE 6.8. The 74138 Octal decoder. Logic diagram and functional Truth Table.

In Figure 6.8, in addition to all the A_i inputs shown in the truth table, the 74138 IC has three additional inputs. These three additional inputs are shown in the logic diagram. These inputs are known as the enable inputs. They function as the gate to the IC. You have to have these inputs at the specified level for the IC to function. These inputs have to have a logic Low level on inputs E_1 and on E_2 while the input E_3 has to have a logic High level ($E_1E_2E_3 \rightarrow 001$).

Historically, this decoder was used for address decoding in a microprocessor system. When a microprocessor puts out a specific address to access the memory system, we have to make sure that only one memory is chosen to provide the response. The decoder uses the address as its input, and the outputs from the decoder are used to enable (alert) one particular memory IC that it is supposed to respond. All the other memory ICs are not enabled

A seven segment display

Lighting up the proper segments to form numbers.

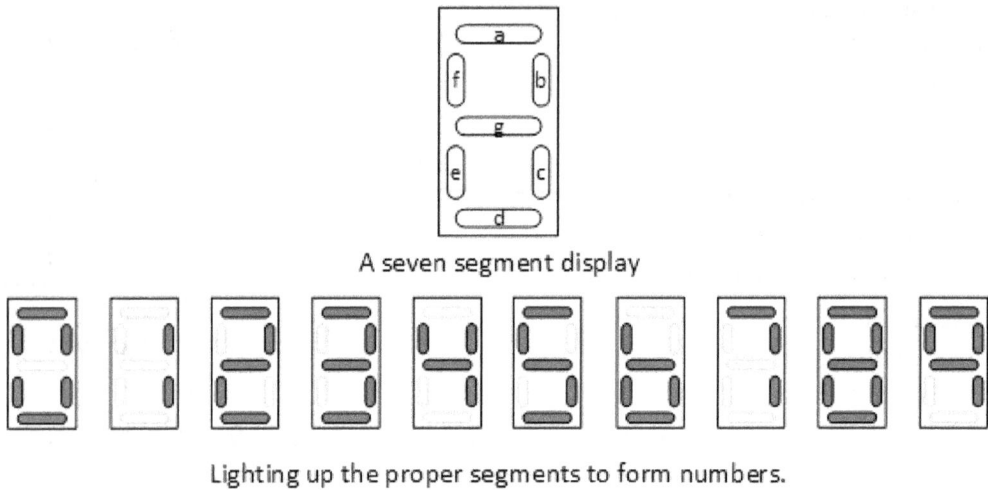

FIGURE 6.9. Seven segment display. How the different numbers are displayed.

so they do not respond. This example is shown in Section 6.3.2 at the end of this chapter.

Another example of a decoder would be BCD to 7-segment decoder. This decoder takes as input the BCD code and converts the signals that would be required by a 7-segment display. Calculators and other numeric displays use the 7-segment display to display numbers by lighting up particular segments of the display. A 7-segment display is shown in Figure 6.9. On the 7-segment display, we identify each of the segments by letters, as shown on the top of Figure 6.9. The bottom of the figure shows which segments we need to light up to display the various numbers from 0 to 9. For example, to display the number 7, we need to light up the segments a, b, and c; to display the number 2, we need to light up the segments a, b, d, e, and g. So the function of the BCD to 7-segment decoder is to treat the BCD number as input and then determine which segments have to be lit up. This truth table is given in Figure 6.10.

The 7447 is the most popular BCD input to the 7-segment decoder IC that converts the BCD input numbers to the 7-segment signals as output. The outputs are active low on the 7447, so when any segment output is low from the decoder IC, the corresponding segment is lit up. As the active output from the 7447 is low, you must use a *Common Anode* 7-segment display

Truth Table for BCD to 7-segment Decoder.

Input				Output						
A_3	A_2	A_1	A_0	a	b	c	d	e	f	g
0	0	0	0	0	0	0	0	0	0	1
0	0	0	1	1	0	0	1	1	1	1
0	0	1	0	0	0	1	0	0	1	0
0	0	1	1	0	0	0	0	1	1	0
0	1	0	0	1	0	0	1	1	0	0
0	1	0	1	0	1	0	0	1	0	0
0	1	1	0	0	0	1	1	1	1	1
0	1	1	1	0	0	0	0	1	1	1
1	0	0	0	0	0	0	0	0	0	0
1	0	0	1	0	0	0	1	1	0	0

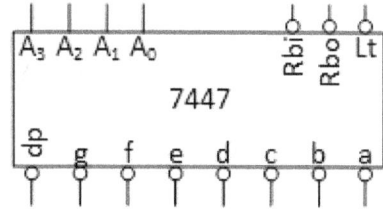

A_3 A_2 A_1 A_0 Rbi Rbo Lt

7447

dp   g   f   e   d   c   b   a

FIGURE 6.10. The BCD to 7-segment decoder. Truth Table and Pin Diagram.

to see the numbers being displayed. Notice that this IC has several other inputs. One of these inputs is the Lt Input. The Lt Input is a *lamp test* input. When this input is active, all the segments will be lit up. This input provides a test function to see if the 7-segment display is functioning properly and none of the segments are damaged. The other inputs are the *Ripple Blanking Input* and the *Ripple Blanking Output*. The purpose of these two pins is to blank out the leading zeros in a multiple digit display, as they are annoying to some people.

6.1.4. ENCODERS

Encoding is the opposite process of decoding. Therefore, an encoded output is in some form of a code; say, the output is in BCD code from a decimal input. Figure 6.11 shows us a couple of possible encoders. They are the decimal to BCD output encoder and the priority encoder.

To build an encoder using combinational logic gates, we begin with the required truth table for the function to be performed. In this truth table, we are treating each of the outputs as a separate function that has to be built but remembering that all the outputs have the same inputs.

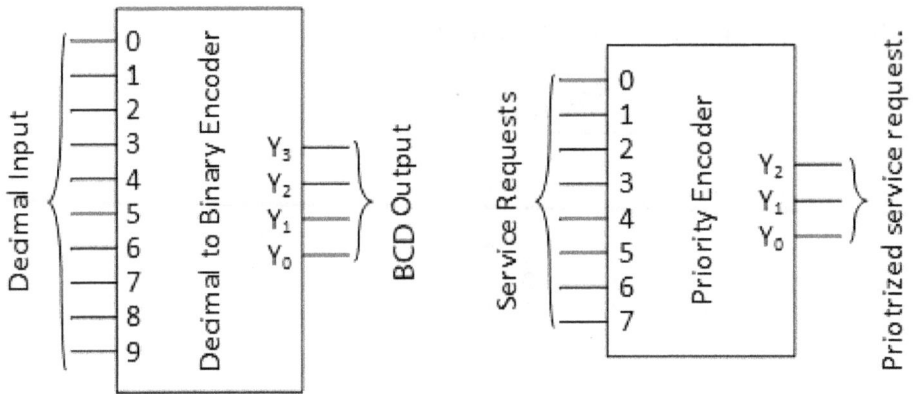

FIGURE 6.11. Pin diagram for a Binary to BCD Encoder and a Priority Encoder.

Here, we will examine the priority encoder. The function of the priority encoder is to identify the highest priority input from all the inputs that are present at that instant. To assign priorities to the inputs, we will choose input 0 as the lowest priority input and input 7 as the highest priority input. With this assignment, we can build a truth table for the priority encoder, as shown in Figure 6.12.

Truth Table for a Priority Encoder.

Input								Output		
0	1	2	3	4	5	6	7	A_2	A_1	A_0
X	X	X	X	X	X	X	L	L	L	L
X	X	X	X	X	X	L	H	L	L	H
X	X	X	X	X	L	H	H	L	H	L
X	X	X	X	L	H	H	H	L	H	H
X	X	X	L	H	H	H	H	H	L	L
X	X	L	H	H	H	H	H	H	L	H
X	L	H	H	H	H	H	H	H	H	L
L	H	H	H	H	H	H	H	H	H	H

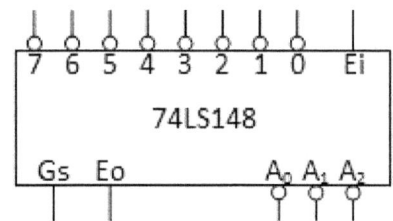

FIGURE 6.12. The Priority Encoder 74LS147. Truth Table and Pin Diagram.

The first thing that we see in Figure 6.12 is that a logic Low input is the active input and the logic Low output is the active output. The first row in the truth table tells us that when input seven is active, all the other inputs are don't care inputs. Therefore, the output indicates that the priority level is level seven by having low on all three outputs. This is the highest level priority. Looking at the second row in the truth table, we see that input seven is inactive, as the level on this input is high. Input six is active, as the level on this input is low. The logic level on any of the other inputs does not matter, as level six is the highest priority level if level seven is not active. Hence, the output indicates this level by setting $A_2A_1A_0$ to $(0\ 0\ 1)_2$. (Remember that in this IC the active level is logic Low.) In the same way, we can determine the priority levels indicated by each of the other rows of the truth table. From the truth table, we can write the logic expression for the three outputs, as shown in Equation 6.4.

$$
\begin{aligned}
A_2 &= 4 \bullet 5 \bullet 6 \bullet 7 \bullet \overline{3} + 3 \bullet 4 \bullet 5 \bullet 6 \bullet 7 \bullet \overline{2} + 2 \bullet 3 \bullet 4 \bullet 5 \bullet 6 \bullet 7 \bullet \overline{1} \\
&\quad + 1 \bullet 2 \bullet 3 \bullet 4 \bullet 5 \bullet 6 \bullet 7 \bullet \overline{0} \\
&= 4 \bullet 5 \bullet 6 \bullet 7 \bullet \overline{3} + 4 \bullet 5 \bullet 6 \bullet 7 \bullet \overline{2} + 4 \bullet 5 \bullet 6 \bullet 7 \bullet \overline{1} + 4 \bullet 5 \bullet 6 \bullet 7 \bullet \overline{0} \\
A_1 &= 6 \bullet 7 \bullet \overline{5} + 5 \bullet 6 \bullet 7 \bullet \overline{4} + 2 \bullet 3 \bullet 4 \bullet 5 \bullet 6 \bullet 7 \bullet \overline{1} + 1 \bullet 2 \bullet 3 \bullet 4 \bullet 5 \bullet 6 \bullet 7 \bullet \overline{0} \\
&= 6 \bullet 7 \bullet \overline{5} + 6 \bullet 7 \bullet \overline{4} + 2 \bullet 3 \bullet 4 \bullet 5 \bullet 6 \bullet 7 \bullet \overline{1} + 2 \bullet 3 \bullet 4 \bullet 5 \bullet 6 \bullet 7 \bullet \overline{0} \\
A_0 &= 7 \bullet \overline{6} + 5 \bullet 6 \bullet 7 \bullet \overline{4} + 3 \bullet 4 \bullet 5 \bullet 6 \bullet 7 \bullet \overline{2} + 1 \bullet 2 \bullet 3 \bullet 4 \bullet 5 \bullet 6 \bullet 7 \bullet \overline{0} \\
&= 7 \bullet \overline{6} + 5 \bullet 7 \bullet \overline{4} + 3 \bullet 4 \bullet 5 \bullet 6 \bullet 7 \bullet \overline{2} + 1 \bullet 3 \bullet 4 \bullet 5 \bullet 6 \bullet 7 \bullet \overline{0}
\end{aligned}
$$

$$(6.4)$$

In writing Equation 6.4, we have made use of logic simplification using Boolean Algebra, according to steps shown in Equation 6.5.

$$
\begin{aligned}
f &= x \bullet \overline{y} + x \bullet y \bullet \overline{z} \\
f &= x \bullet \overline{y} \bullet z + x \bullet \overline{y} \bullet \overline{z} + x \bullet y \bullet \overline{z} \\
f &= x \bullet \overline{y} \bullet (z + \overline{z}) + x \bullet \overline{z} \bullet (y + \overline{y}) \\
f &= x \bullet \overline{y} + x \bullet \overline{z}
\end{aligned}
$$

$$(6.5)$$

This exercise shows us that building a logic function—no matter what the function is supposed to accomplish—is simply a matter of writing out the truth table first. Next, from the truth table, we write the logic equation. This logic equation is simplified so that it is a minimum cost function. The minimum cost function is then used to build the logic function. It is this simple idea (along with the knowledge that all two-level logic functions can be built using the AND-OR-INVERT logic) that is used to build functions using Programmable Array of Logic Gates.

Question: Develop a truth table for the decimal to BCD encoder.
  Answer: The truth table is shown here.

Question: Write the logic equation for output A_2.
  Answer: Output A_2 will have the following logic equation.

$$A_2 = 7 \bullet \overline{8} \bullet \overline{9} + 6 \bullet \overline{7} \bullet \overline{8} \bullet \overline{9} + 5 \bullet \overline{6} \bullet \overline{7} \bullet \overline{8} \bullet \overline{9} + 4 \bullet \overline{5} \bullet \overline{6} \bullet \overline{7} \bullet \overline{8} \bullet \overline{9}$$

$$A_2 = 7 \bullet \overline{8} \bullet \overline{9} + 6 \bullet \overline{8} \bullet \overline{9} + 5 \bullet \overline{8} \bullet \overline{9} + 4 \bullet \overline{8} \bullet \overline{9}$$

6.2.   PROGRAMMABLE ARRAY OF LOGIC GATES

We have seen that the cost to build a logic function is very strongly dependent on the component count. Reducing the component count plays a very important role in the cost of building a logic function. This cost of component count applies equally to individual gates and to entire ICs, so it would be very beneficial if we were able to reduce not only the number of gates, but the number of ICs as well. Toward this end, earlier in this chapter, we saw some standard logic functions and how they help in reducing the component count by including the entire function in one IC. We can extend this idea much further if we give the designer a flexible AND-OR-INVERT structure in one IC where the designer can choose the inputs to the AND gate and then choose the inputs to the OR gate. That is specifically the idea behind the programmable array of Logic gates.

How do we give the designer the choice to select the inputs to the AND gates and the OR gates? The answer is that we first connect all the inputs to each of the AND gates (along with their complements) and then connect the

Truth Table for a Decimal to BCD Encoder.

Input										Output			
0	1	2	3	4	5	6	7	8	9	A_3	A_2	A_1	A_0
X	X	X	X	X	X	X	X	X	1	1	0	0	1
X	X	X	X	X	X	X	X	1	0	1	0	0	0
X	X	X	X	X	X	X	1	0	0	0	1	1	1
X	X	X	X	X	X	1	0	0	0	0	1	1	0
X	X	X	X	X	1	0	0	0	0	0	1	0	1
X	X	X	X	1	0	0	0	0	0	0	1	0	0
X	X	X	1	0	0	0	0	0	0	0	0	1	1
X	X	1	0	0	0	0	0	0	0	0	0	1	0
X	1	0	0	0	0	0	0	0	0	0	0	0	1
1	0	0	0	0	0	0	0	0	0	0	0	0	0

Truth Table for Decimal to BCD Encoder.

outputs of all the AND gates to the input of each of the OR gates. Then we let designers selectively disconnect the inputs to the AND gates that they do not need. In the same way, we allow the designers to selectively disconnect the inputs to the OR gates that they do not need. Following this procedure, the designer can custom design a programmable array of logic gates to suit the function that he wants to build. Such general-purpose logic building blocks are known as *Programmable Array Logic* (PAL) or *Programmable Logic Array* (PLA).

6.2.1. WHAT ARE PALS AND PLAS?

Figure 6.13 shows a general block diagram of a component that is a programmable array logic. These devices have multiple inputs and multiple outputs. All the inputs generally enter into an array of AND gates. All the inputs and their complements are inputs to each of the AND gates. The

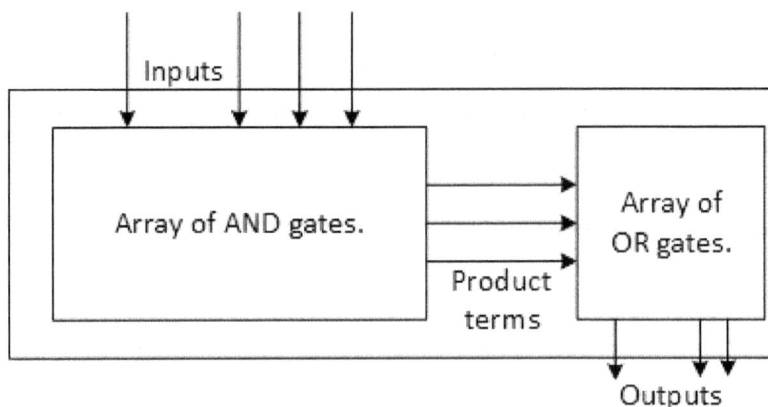

FIGURE 6.13. Organization of a Programmable Logic Array.

AND gates are used to develop the product terms of the required function. The output of the AND gates make the product terms. The product terms are then used as inputs to the OR array. The output of all the AND gates are inputs to each of the OR gates. The OR array combines the product terms to complete the sum of products function. There are devices that have both the arrays programmable or just the AND array programmable while the OR array connections are fixed.

When both the AND array and the OR array are programmable, we call the device a *Programmable Logic Array* (**PLA**). Not all devices have full programmability; some devices will have programmable AND arrays, but once the product terms are formed, they are routed to specific OR gates. The connections from the AND gates to the OR gates are preset at the factory, and they are not programmable by the user. The devices that have partial programmability are known as *Programmable Array Logic* (**PAL**).

There is a tradeoff in PAL devices between the complexity of the function in terms of the product terms per OR gate and the number of independent functions that the device can build. When there is a high number of fan-ins to the OR gates (outputs from many AND gates connected to the input of one OR gate), the function built has many product terms. Such functions are more complex and only a few can be built in one device. When there are many OR gates, there are fewer connections from the AND gates into each OR gate and a low number of fan-ins. Such functions are generally less complex and we can build many different functions in one device. In Figure 6.14, we see an un-programmed PLA device. This is a PLA device as both the AND array and the OR array are programmable.

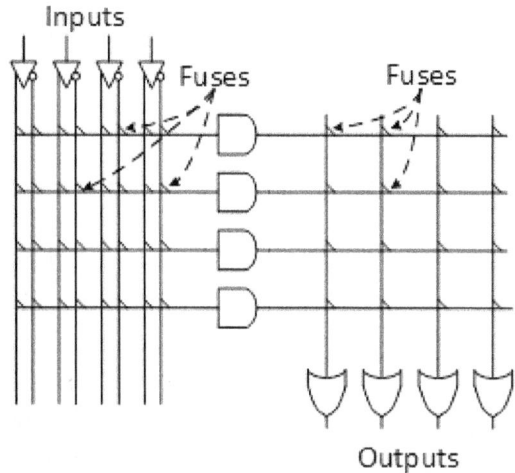

FIGURE 6.14. A 4 input 4 output PLA device.

The first thing that you see in Figure 6.14 is that all the inputs and the complements of all the input variables are connected to each of the AND gates. If we were to leave the connections as they are, then the output of all the AND gates would be zero always. To make the AND gates perform a logic function, we must remove some of the connections. This is what we mean by programming the device. To program the device, we must blow the unwanted fuses and leave the remaining fuses intact. When we do this correctly, we will form the product terms as outputs of the AND gates.

Once the AND array is programmed, we can program the OR array. To program the OR array, we select the outputs of the AND gates that are part of one logic function. We leave these fuses intact while we blow the other fuses to the OR gate. This way, we can build several different logic functions in one IC.

6.2.2. USING THE PROGRAMMABLE ARRAY OF LOGIC GATES

A PLA device can be used to implement a logic function of a high level of complexity. The complexity of the logic function is determined by the number of inputs to the PLA device, the number of product terms required (the number of AND gates used), and the number of different functions the device has to implement (the number of OR gates). Suppose you want to implement all four Boolean functions given in Equation 6.6.

$$F_1 = A + \overline{B} \bullet \overline{C} \qquad\qquad F_2 = A \bullet \overline{C} + \overline{B} \bullet \overline{C}$$

$$F_3 = A \bullet B + \overline{B} \bullet \overline{C} \qquad\qquad F_4 = A \bullet \overline{C} + A \bullet B \qquad (6.6)$$

We note that to build the four functions, we need a PLA device that has at least three different inputs; four different AND gates, as we have four unique product terms $(A; \ \overline{B} \bullet \overline{C}; \ A \bullet \overline{C}; \ A \bullet B)$; and four OR gates, as we have four different functions that we have to build. All four functions can be built in a PLA like the one shown in Figure 6.14. Figure 6.15 shows how we would build the four functions given in Equation 6.6.

In Figure 6.15, you should notice that we have burned the fuses that we do not want. All the other fuses are left intact. This way, the four AND gates build the four product terms and the outputs from the product terms are gathered by the OR gates to build the four logic functions.

Question: This time, we want to build the following functions using a PLA device. Show which fuses you will burn to build the required functions.

$$F_1 = A \oplus B \oplus C \qquad\qquad F_2 = A \bullet \overline{B} \bullet C + A \bullet B \bullet \overline{C} + \overline{A} \bullet B \bullet C$$

$$F_3 = A \bullet B \qquad\qquad F_4 = A \bullet \overline{B} \bullet C + \overline{A} \bullet B \bullet C$$

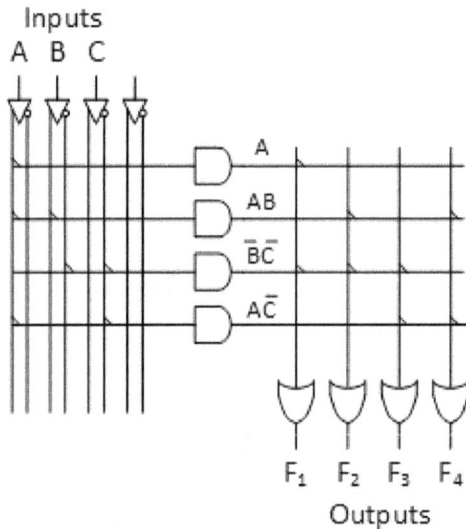

FIGURE 6.15. A PLA device implementing 4 functions of 3 inputs.

Answer: We can build this function using the PLA device shown in Figure 6.14 if it needs only four different product terms. It appears that all the functions together need more than four product terms. However, if we rewrite the four functions as shown in Equation 6.7, we see that only four product terms are required to build all four functions.

$$F_1 = A \bullet \bar{B} \bullet C + A \bullet B \bullet \bar{C} + \bar{A} \bullet B \bullet C + A \bullet B \bullet C$$
$$F_2 = A \bullet \bar{B} \bullet C + A \bullet B \bullet \bar{C} + \bar{A} \bullet B \bullet C$$
$$F_3 = A \bullet B \bullet \bar{C} + A \bullet B \bullet C$$
$$F_4 = A \bullet \bar{B} \bullet C + \bar{A} \bullet B \bullet C$$

$$(6.7)$$

When we write all the functions as shown in Equation 6.7, we see that there are only four unique product terms and we can build all the functions using the PLA device shown in Figure 6.14.

6.3.   EXAMPLES OF COMBINATIONAL LOGIC

In this section, we examine two different design examples. One is a code converter and the other is an example of address decoding in a microprocessor. Although both the design examples could be completed either by a PLA device or by using logic function ICs like the decoder and the multiplexer, we will complete one design example, the code converter, using the PLA device and the other example, microprocessor decoding, using the decoder device. The examples will highlight the various alternatives that an engineer has to implement combinational logic.

6.3.1.   A BINARY TO GRAY CODE CONVERTER

In this example, we will build a binary to Gray code converter. In Figure 6.16, we see the truth table and the K-maps for the binary to Gray code converter.

Inputs				Outputs			
W	X	Y	Z	G_3	G_2	G_1	G_0
0	0	0	0	0	0	0	0
0	0	0	1	0	0	0	1
0	0	1	0	0	0	1	1
0	0	1	1	0	0	1	0
0	1	0	0	0	1	1	0
0	1	0	1	0	1	1	1
0	1	1	0	0	1	0	1
0	1	1	1	0	1	0	0
1	0	0	0	1	1	0	0
1	0	0	1	1	1	0	1
1	0	1	0	1	1	1	1
1	0	1	1	1	1	1	0
1	1	0	0	1	0	1	0
1	1	0	1	1	0	1	1
1	1	1	0	1	0	0	1
1	1	1	1	1	0	0	0

K-Map for G_3:

YZ \ WX	00	01	11	10
00	0	0	1	1
01	0	0	1	1
11	0	0	1	1
10	0	0	1	1

$G_3 = W$

K-Map for G_2:

YZ \ WX	00	01	11	10
00	0	1	0	1
01	0	1	0	1
11	0	1	0	1
10	0	1	0	1

$G_2 = \overline{W}X + W\overline{X}$

K-Map for G_1:

YZ \ WX	00	01	11	10
00	0	1	1	0
01	0	1	1	0
11	1	0	0	1
10	1	0	0	1

$G_1 = \overline{X}Y + X\overline{Y}$

K-Map for G_0:

YZ \ WX	00	01	11	10
00	0	0	0	0
01	1	1	1	1
11	0	0	0	0
10	1	1	1	1

$G_0 = \overline{Y}Z + Y\overline{Z}$

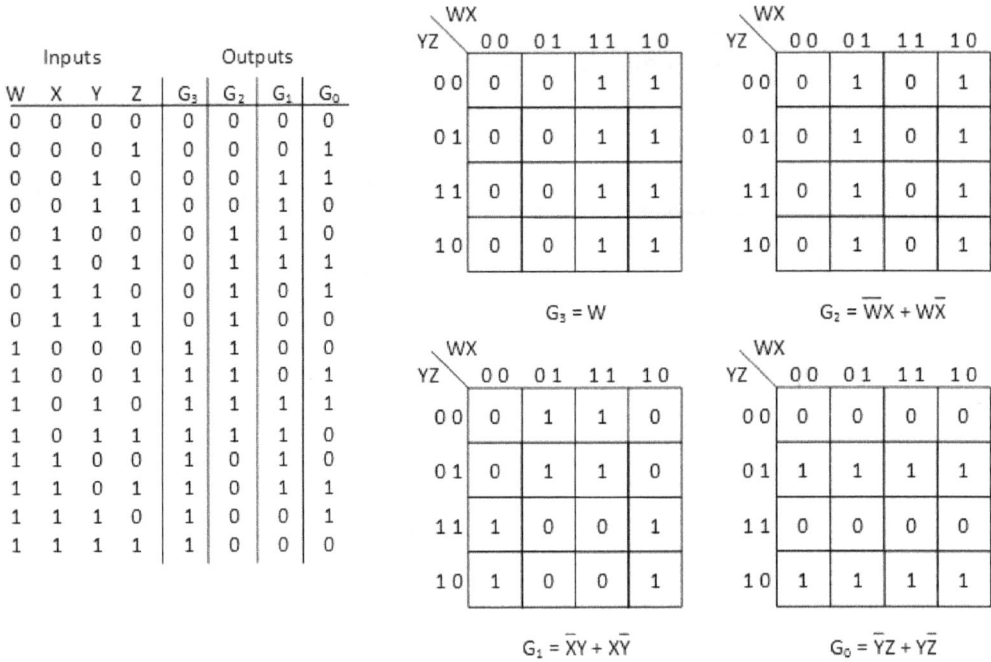

FIGURE 6.16. Truth table and K-Map for Binary to Gray code converter.

Figure 6.17 shows us the PLA connection to implement the binary to Gray code conversion. The four logic functions that we have to build are given in Equation 6.8. This time, we are not able to use any of the product terms again since none are common. As a result, we need seven AND gates, and as there are four outputs, we need four OR gates to build the function. It is interesting to note that if we used discrete logic, we would need six AND gates and three OR gates.

$$
\begin{aligned}
G_3 &= W \\
G_2 &= \overline{W} \bullet X + W \bullet \overline{X} \\
G_1 &= \overline{X} \bullet Y + X \bullet \overline{Y} \\
G_0 &= \overline{Y} \bullet Z + Y \bullet \overline{Z}
\end{aligned}
\tag{6.8}
$$

Each of these gates are two-input gates, so using discrete logic, we could have built this code converter using three ICs (Two 7408 Quad AND gates and one 7432 Quad OR gate.) Using a PLA, we are able to build this converter using only one IC—a definite saving.

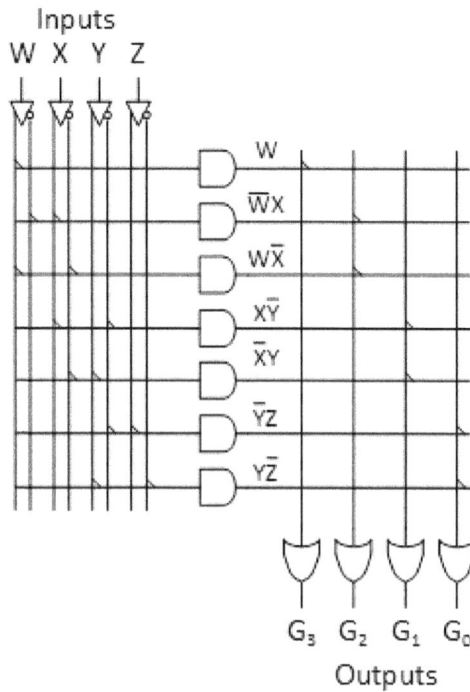

FIGURE 6.17. A PLA device implementing the Binary to Gray Code Converter.

6.3.2. MICROPROCESSOR ADDRESS DECODING

Microprocessors are required to locate thousands of different data items to function correctly. To locate a data item, the microprocessor uses an address. An address in a microprocessor system is a binary number. When the microprocessor puts out an address, the system has to use the address to alert the data item that it is being requested by the microprocessor. This is known as decoding the address.

The data items requested by the microprocessor are generally stored in memories. In any microprocessor system, there will be several memory devices. Therefore, to select the requested data item, we have to first select the memory device that holds the data item and then bring the requested data item out from within the memory device and present it to the microprocessor.

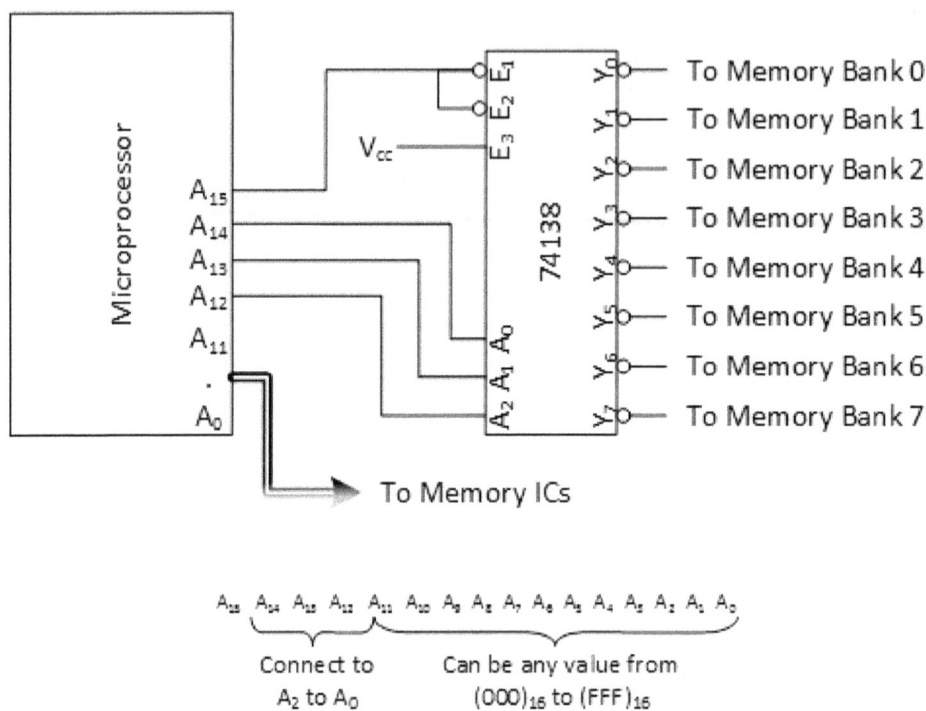

FIGURE 6.18. Address decoding for a microprocessor system.

A typical eight-bit microprocessor like the Intel 8085 or the Motorola 6800 can address 2^{16} different addresses. To address all these memory locations, the microprocessor uses sixteen address lines (A_{15} to A_0). Out of these sixteen address lines, the low order address lines are used by the memory itself to select the data word from all the data words stored in the memory. Therefore, for example, address lines A_{11} to A_0 are directly connected to the memory. The other address lines are used by the address decoding system to select the memory device. This arrangement is shown in Figure 6.18.

In Figure 6.18, we see sixteen address lines out of the microprocessor. Address line A_{15} is connected to the enable inputs and must be logic Low. When address line A_{15} is low, the decoder shown will be selected. Next, the address lines A_{14}, A_{13}, and A_{12} are connected to the decoder inputs A_2, A_1, and A_0. With this connection, memory bank 0 will be selected when the address is $(0\ 0\ 0\ 0\ x \ldots x)_2$, which is addresses from 0_{10} to 4095_{10}. Similarly, we can determine the addresses for the other memory banks.

Question: A building alarm system has detectors to determine if secure doors are left open. There are eight doors that are sensitive. The security person would like to identify which sensitive door is left open. Design a system that will inform the security person which door is left open.

   Answer: A system that will identify the sensitive door left open can be built with a priority encoder. This is shown in Figure 6.19. The sensor output from the eight doors is connected to the eight inputs of the 74148. The Ei input that enables the priority encoder is permanently connected to logic Low so the priority encoder is always checking the door sensors. The output will be from the door with the highest priority first, and when that particular input is not active, the output will indicate the next door that is open.

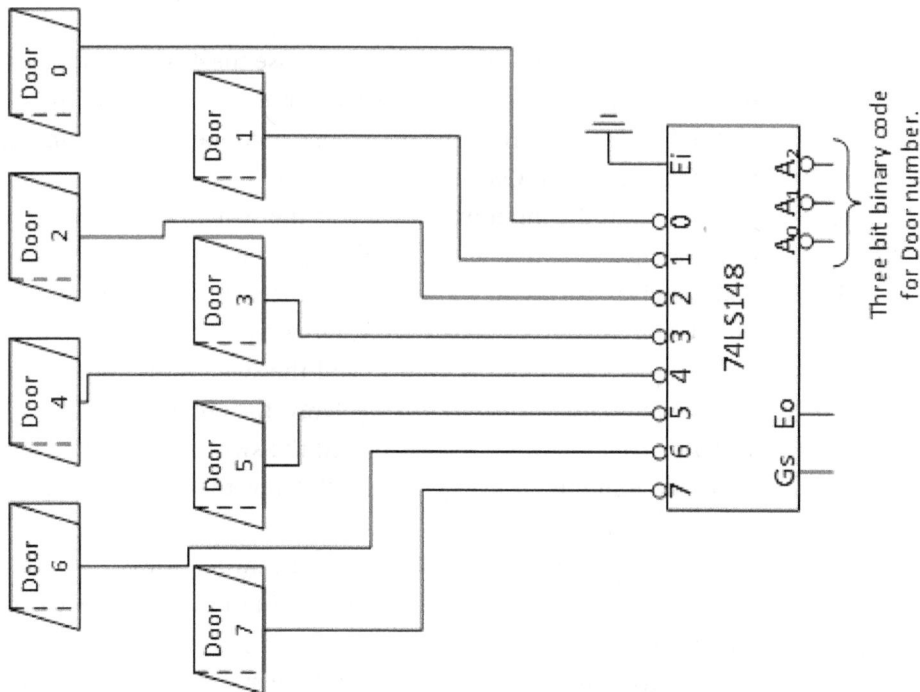

FIGURE 6.19. Using the Priority Encoder 74LS148 to identify open doors.

6.4. CHAPTER PROBLEMS

6.4.1. You are given the following function of four variables. Use Shannon's Expansion Theorem to expand the function about each of four variables individually.

$$f_{abcd} = \sum m_0, m_1, m_4, m_6, m_9, m_{11}, m_{14}, m_{15}$$

6.4.2. For the function given in Problem 6.4.1, use the 4-to-1 multiplexer to build the function when the switch variables $S_1 S_0$ are variables A and C.

6.4.3. For the function given in Problem 6.4.1, use the 4-to-1 multiplexer to build the function when the switch variables $S_1 S_0$ are variables B and C.

6.4.4. For the function given in Problem 6.4.1, use the 4-to-1 multiplexer to build the function when the switch variables $S_1 S_0$ are variables D and C.

6.4.5. For the function given in Problem 6.4.1, use the 4-to-1 multiplexer to build the function when the switch variables $S_1 S_0$ are variables C and B. Is there a difference between what is connected to the inputs between Q6.4.3 and Q6.4.5?

6.4.6. You are given the input waveform and the switch waveforms for a 4-to-1 multiplexer in Figure Q6.4.6. Draw the output waveform.

6.4.7. You are given the input waveform and the switch waveforms for a 4-to-1 multiplexer in Figure Q6.4.7. Draw the output waveform.

6.4.8. You are given the input waveform and the switch waveforms for a 4-to-1 multiplexer in Figure Q6.4.8. Draw the output waveform.

6.4.9. You are given the input waveform and the switch waveforms for a 1-to-4 de-multiplexer in Figure Q6.4.9. Draw the output waveform for each of the four outputs.

6.4.10. You are given the input waveform and the switch waveforms for a 1-to-4 de-multiplexer in Figure Q6.4.10. Draw the output waveform for each of the four outputs.

6.4.11. You are given the input waveform and the switch waveforms for a 1-to-4 de-multiplexer in Figure Q6.4.11. Draw the output waveform for each of the four outputs.

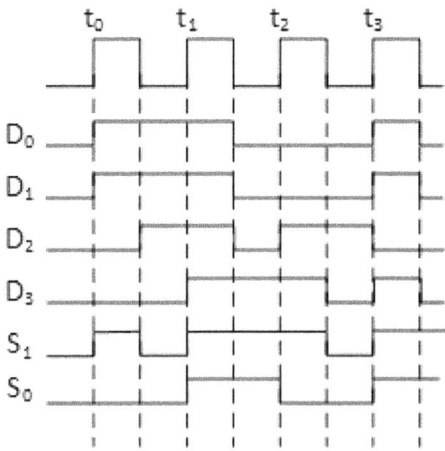

FIGURE Q6.4.6. Input signals and the switch waveforms.

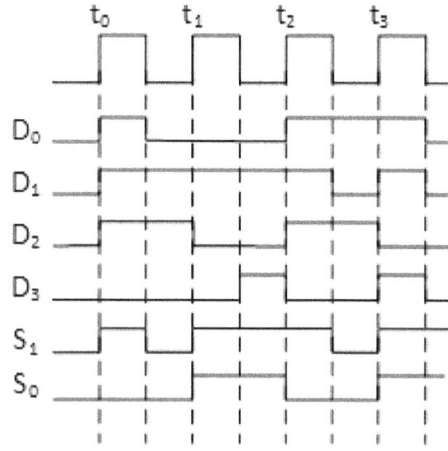

FIGURE Q6.4.7. Input signals and the switch waveforms.

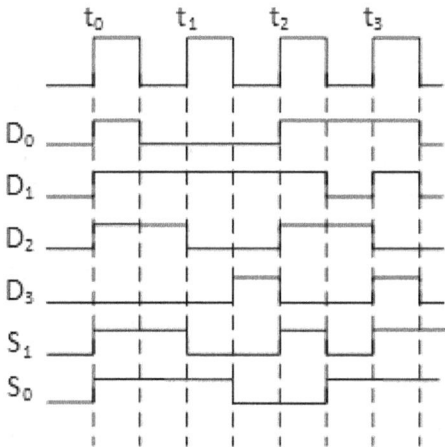

FIGURE Q6.4.8. Input signals and the switch waveforms.

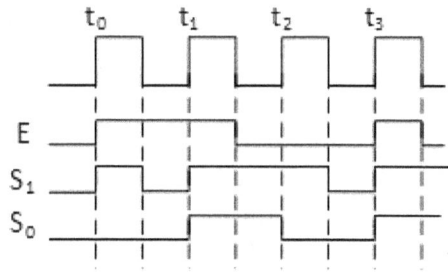

FIGURE Q6.4.9. Input signals and the switch waveforms.

6.4.12. An Excess-3 code is formed by increasing the binary code by three, so a zero in Excess-3 code will be written as $(0011)_2$. Write the Excess-3 code for decimal digits 0 to 9. Next, write the truth table that will decode the Excess-3 code to a BCD code.

6.4.13. Using the truth table from Q6.4.12, write the logic expression for each of the BCD digits.

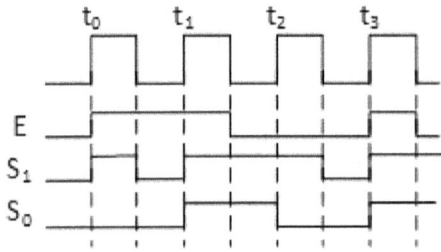

FIGURE Q6.4.10. Input signals and the switch waveforms.

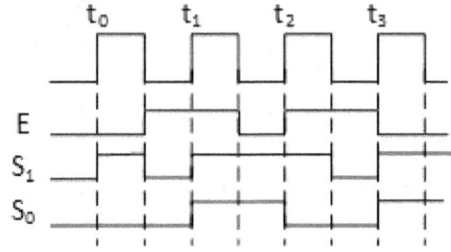

FIGURE Q6.4.11. Input signals and the switch waveforms.

6.4.14. The IC 7447 is a BCD to 7-segment decoder. Write the truth table for Excess-3 to 7-segment decoder.

6.4.15. The 74LS148 is a priority encoder that encodes inputs to straight binary. For this question, write a truth table that can be used to design a priority encoder that will encode eight inputs to Excess-3 code.

6.4.16. From the truth table in Question 6.4.15, determine the logic expressions for the output.

6.4.17. What are the differences between the PLA and the PAL devices?

6.4.18. All logic functions can be built using either a PLA or a PAL. Here, we want to determine the required connections to use a PLA like the one shown in Figure 6.14 to build the decoder of Question 6.4.14. Draw the required diagram. If you need more AND gates you may use them.

6.4.19. This time, we want to determine the required connections to use a PLA like the one shown in Figure 6.14 to build an encoder of Question 6.4.15. Draw the required diagram. If you need more AND gates you may use them.

7 FLIP-FLOPS AND COUNTERS

7.0. INTRODUCTION

In the first six chapters we have examined circuits that have used simple logic gates, such as the AND, OR, and NOT gates. The characteristics of such circuits are that the output from the circuit is strictly a combinational output that depends only on the current inputs. These circuits are instantaneous. The output is obtained after a few (typically two) gate delays. There is no time component involved in the circuits. The circuits also had no memory; hence, the present output is independent of any and all past outputs and inputs. In this chapter, we embark on circuits that use the present output as part of its next input. Such a circuit will need to store the history of past inputs so these circuits have memory. The circuit also has a timing component, as the inputs need to be recognized at a specific instant in time and not before. To study the circuits that have memory, we will begin our study with an element that can hold one bit of information in its memory. The information is stored until the new input is recognized, which may change the information stored in the circuit. These circuits are then used in the next chapter to build circuits that are more complex.

7.1. CIRCUITS WITH FEEDBACK

All of us have seen a digital clock. The present output of the clock, which is the present time, is updated every second. To update the time and show it on the output, the clock has to remember what the previous output was. The clock has to wait for the next timing signal (a timing signal that tells the clock that one second has elapsed) to arrive. When the timing signal arrives, the clock will take the present output, update it, and give us the next output. In this example, the present output is treated as an input. Since the output is treated as an input, this builds a feedback in the circuit. The signal that tells the clock that it has to update the output (the timing signal) is provided by some kind of an oscillator. Often this oscillator is a crystal oscillating at a known frequency. By dividing the known frequency of the crystal oscillator, we can get a signal or a pulse, once every second, to update the time on the clock. The clock circuit simply counts these pulses once every second to keep track of the time. So the clock takes its present output and the next input to give us the next output. Interestingly, the timing pulse that keeps our clock displaying the correct time is known as the *Clock* signal.

Consider a different system: this time we will try to see what happens inside a vending machine. Every time you deposit a coin, the display shows how much money you have deposited. Initially, the circuit is counting the money deposited. Once you have deposited enough money to purchase an item, the vending machine switches states and looks for the input that indicates the item you want the machine to give to you. When you enter the number of the item, the machine will recognize the item number. The machine will now switch states so it can dispense the item to you. This machine has to remember that you have entered enough money to purchase an item, remember which item you have chosen, and, finally, dispense the item. All these circuits have two things in common: they have memory and they follow a sequence. We will examine the memory portion of these circuits in this chapter and in the next chapter; we will see how we make these circuits follow a required sequence of steps.

7.1.1.   SIMPLE CIRCUITS WITH FEEDBACK

Examine the circuit shown in Figure 7.1. In this circuit, the output of one inverter is the input to the second inverter, and the output of the second inverter is the input to the first inverter. This way, the output of the first inverter is reinforced by having the complement of the output as the input. This reinforcement makes the first inverter remember the output. If a 0 is entered as the input to the first inverter, then the circuit will remember the output of the first inverter as 1 until a different value is entered to the input of the first inverter. This is specifically the problem with this circuit: how do you enter a new value as the input to the first inverter in this circuit? The circuit will have some value when we apply power to the circuit, but how do we make this value a specific value that we want? To get a specific value that we want, we will have to introduce some other logic so that we can have the circuit remember the last value entered; at the same time, we should be able to change that value whenever we want.

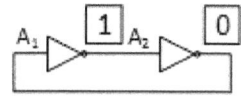

FIGURE 7.1. A very simple Memory Circuit

In Figure 7.1, we had an even number of inverters in cascade. Consider now the circuit shown in Figure 7.2. In that circuit, we have an odd number of inverters in a chain. This time, suppose we have been able to enter a logic low as the input for inverter A_1, which gives us a logic high as the output from inverter A_1. This output will go through the chain of inverters, and when the signal comes back to the input of inverter A_1, it will be logic high. This will change the output of inverter A_1 to low, and this will go around the chain. When this comes back to the input of inverter A_1, it will be a logic low. With this input, the output of inverter A_1 will change to a logic high output. With this chain of an odd number of inverters, we see an oscillatory behavior when we examine the output of any of the inverters in the chain.

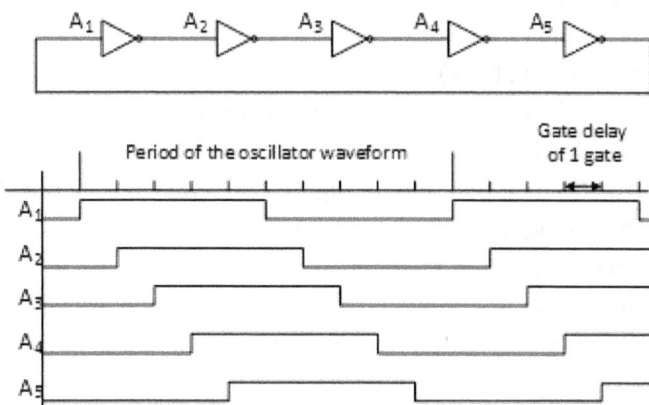

FIGURE 7.2. Chain of Inverters and Timing Diagram.

7.1.2. CROSS-COUPLED NOR GATES

We now go back to a circuit like the one we saw in Figure 7.1. This time, instead of using inverters, we have used NOR gates (as shown in Figure 7.3). The circuit is shown in two different layouts. The first is the standard form that is used to represent the circuit. The second is the circuit that represents the diagram (very similar to the circuit in Figure 7.1). Let us evaluate and see how the circuit works. In this circuit, we are using NOR gates. To review how NOR gates behave, remember that, with NOR gates, when one of the inputs is a logic high, then the output from the NOR gate will always be logic low; when one of the inputs is a logic low, then the NOR gate acts like an inverter with respect to the other input. In this case, the output will always be the complement of the other input.

Now suppose the R-S inputs to the two NOR gates in Figure 7.3 are $(0\ 1)_2$. Since the S input to NOR gate 2 is high, the output of that gate, Q, will be logic low. This logic low output of NOR gate 2 is the input to NOR gate 1. With a low input, this gate behaves like an inverter for the other input, which is the R input. Since the R input this time is low, the output of NOR gate 1 will be high. With the R-S inputs being $(0\ 1)_2$, the Q output is high.

Now suppose the R S inputs to the NOR gates are $(1\ 0)_2$. Since the R input to NOR gate 1 is high, the output of that gate, Q, will be logic low. This logic low output of NOR gate 1 is the input to NOR gate 2. With a low input, this gate behaves like an inverter for the other input, which is the S input. Since the S input is low this time, the output of NOR gate 2 will be high. With the R S inputs being $(1\ 0)_2$, the Q output is low.

Finally, when the R S inputs to the NOR gates are $(0\ 0)_2$, then both of the NOR gates will behave like chained inverters (i.e., the circuit is behaving exactly like the chained inverters in Figure 7.1). This permits the circuit to remember the last input. With the cross-coupled NOR gates, we have built a memory circuit that can remember the last input when the R S inputs are $(0\ 0)_2$, and this circuit will allow us to enter the value that we want remembered when the R S inputs are either $(0\ 1)_2$ or $(1\ 0)_2$.

7.1.3. TIMING BEHAVIOR AND TRUTH TABLE OF CROSS-COUPLED NOR GATES

We have seen in the previous section that the output of the circuit in Figure 7.3 behaves as follows: when the inputs (R S) are $(0\ 0)_2$, then the circuit remembers the last input. This input is known as the memory state. When the inputs (R S) are $(0\ 1)_2$, then the Q output of the circuit is set to high. This input is known as the Set state. When the inputs (R S) are $(1\ 0)_2$, then the Q output of the circuit is set to low. This input is known as the Reset state. Finally, when the input (R S) is $(1\ 1)_2$, we see that this particular input is not allowed. This input is not allowed for the following reasons. First, consider the $(1\ 1)_2$ input, identified as ① in Figure 7.3. Here we see that both the outputs from the circuit are logic low. This is not what we want from the circuit. We want the two outputs to be complements of each other. Next, consider the $(1\ 1)_2$ input, identified as ② in Figure 7.3. This is the real reason why we do not permit the input $(1\ 1)_2$ for this circuit. This time, the special condition is that the inputs (R S) changed from $(1\ 1)_2$ to $(0\ 0)_2$. Now the circuit begins to oscillate, and the circuit cannot decide what the output from the circuit should be. This ambiguous behavior is the main reason we do not permit the circuit to have $(1\ 1)_2$ as an input. The circuit oscillations that take place in the output are explained below.

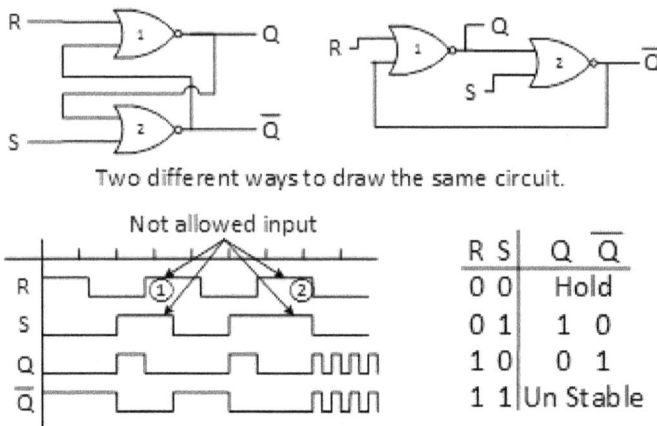

Two different ways to draw the same circuit.

R S	Q	$\overline{Q}$
0 0	Hold	
0 1	1	0
1 0	0	1
1 1	Un Stable	

FIGURE 7.3. Cross-coupled N or Gates. The S-R latch.

Initially, both the Q and the $\overline{Q}$ outputs of the circuit will be set to 0, as both the R and the S inputs were 1. Now, when both the R and the S inputs change to $(0\ 0)_2$ simultaneously, both the NOR gates have $(0\ 0)_2$ inputs. With $(0\ 0)_2$ inputs to the two NOR gates, the output from both the gates is a 1. Therefore, the outputs of both of the NOR gates will change to 1 simultaneously. This output from one NOR gate is also the input to the other NOR gate. Since the input to the NOR gate is a 1, the output from the NOR gate will change to 0. Since both the NOR gates have an input of 1, both of the NOR gate outputs will change to 0 simultaneously. This is how the circuit started. We have come full circle, and we will go around again and again this way until one of the inputs changes from $(0\ 0)_2$ to something else. For this reason, we say that the input to the cross-coupled NOR gates must never be allowed to be $(1\ 1)_2$. The circuit of the cross-coupled NOR gates (like the one shown in Figure 7.3) is known as the S-R latch.

Review Questions for Section 7.1

Question: Draw the circuit of the cross-coupled NOR gates and write the truth table for the circuit.

    Answer: Figure 7.3 shows the circuit and the truth table.

Question: In the circuit of the cross-coupled NOR gates, replace the NOR gates by NAND gates and write the truth table.

    Answer: Figure 7.4 shows the circuit with the NAND gates and the truth table.

Note the two different circuits of the R S latch. You must have noticed that both of the circuits for the R S latch (one with the NOR gates and the

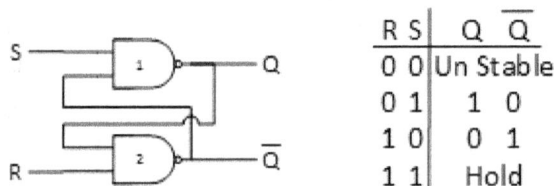

R S	Q	$\overline{Q}$
0 0	Un Stable	
0 1	1	0
1 0	0	1
1 1	Hold	

FIGURE 7.4. Cross-coupled NAND Gates. The S-R latch.

other with the NAND gates), along with their truth tables, are very similar to each other. The truth tables are similar but not exactly the same. We have also seen that with the ICs that we get, we only see the package on the outside. We do not see what is on the inside of the IC. So how can we know if the circuit is built using NOR gates or NAND gates? The good news is that we do not need to know the answer to this question. The engineers who design the circuits have all agreed that the truth table of only the NOR gate circuits will always be followed, no matter how the circuit is designed and built, so we will always consider the $(0\ 0)_2$ input to the R S latch as the memory input and the $(1\ 1)_2$ input to the R S latch as the input that is not allowed. The truth table is always like the one shown in Figure 7.3. This circuit is known as the S-R Latch.

7.2. LATCHES AND FLIP-FLOPS

We have shown that the circuit in Figure 7.3 (the S-R latch) is a memory device that can remember one bit of information. In the S-R latch, the output changes immediately after (following one or two gate delays) the new input is applied. We call this *level-triggered*, meaning that when the level of the input is changed, the change in the output is also triggered. This is sometimes not an acceptable condition. We may want to output to trigger with the occurrence of another event. This implies that we should present the new levels to the S-R latch with the occurrence of this external event. This is done by a *clock* signal. Examine Figure 7.5.

Figure 7.5 has the same R-S latch, but at the input we have added two AND gates, along with a clock signal. With this input configuration, when

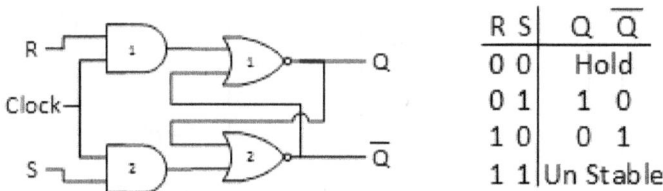

R S	Q $\overline{Q}$
0 0	Hold
0 1	1 0
1 0	0 1
1 1	Un Stable

FIGURE 7.5. The clocked S-R Flip-Flop.

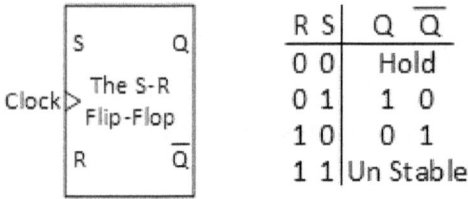

R S	Q	$\overline{Q}$
0 0	Hold	
0 1	1	0
1 0	0	1
1 1	Un Stable	

FIGURE 7.6. The logic diagram of the S-R Flip-flop.

the clock input is low, then the input to the S-R latch is $(0\ 0)_2$, which, as we know, represents a no-change condition. Under this condition, the circuit maintains the previous output. When the clock input is 1, the two AND gates are *enabled*, and the output from the circuit follows the truth table logic of the S-R latch. We distinguish the flow-through operation and the clocked operation of the two circuits by calling the clocked circuit a flip-flop and the flow-through operation a latch. The flip-flop is usually drawn in logic circuits (as shown in Figure 7.6).

We can have several different types of clocking circuits. The one shown in Figure 7.5 is known as a positive level-triggered flip-flop, since the triggering occurs on the rising edge of the clock signal, or when the clock signal is going from logic low to logic high. We call this the *rising-edge-triggered flip-flop* if the triggering occurs only at the instant of the rising edge of the clock, and not when the clock level is high. If we place inverters at the input of the AND gate for the clock signal, then the flip-flop will trigger when the clock signal is going from logic high to logic low. We call this the *falling-edge - triggered flip-flop* if the triggering occurs only at the instant of the falling edge of the clock, and not when the clock level is low. The third clocking method is the *Master—Slave-triggered flip-flop*. This is shown in Figure 7.7.

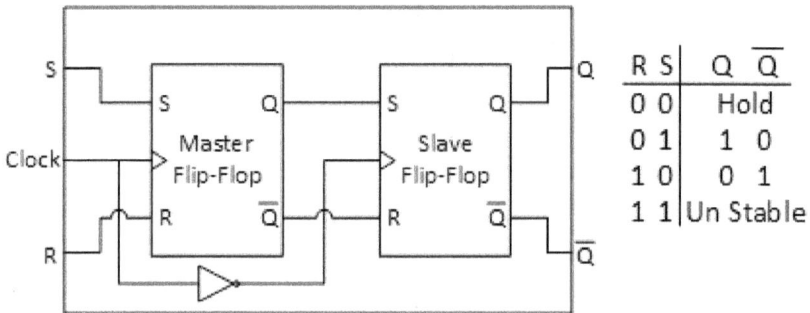

R S	Q	$\overline{Q}$
0 0	Hold	
0 1	1	0
1 0	0	1
1 1	Un Stable	

FIGURE 7.7. The Master—Slave S-R Flip-Flop.

In a Master—Slave flip-flop, we use two flip-flops: the Q and the Q outputs of the first flip-flop are used as the S R inputs to the second flip-flop. The clock input to the second flip-flop is inverted when compared to the clock input to the first flip-flop. With this arrangement, the first (or the master) flip-flop will trigger on one-half of the clock signal. When the clock signal changes state, the first flip-flop will be disabled and the second (or the slave) flip-flop will be enabled. The input to the second flip-flop is the output from the first flip-flop, so the second, or slave, flip-flop will follow the output from the first flip-flop. It is for this reason that this flip-flop is known as the slave flip-flop.

7.2.1. TIMING THE SIGNALS IN FLIP-FLOPS

Signals in all logic circuits need to meet some specific timing requirements to be recognized by the logic gate. We can break this time interval down into the *setup time* and the *hold time*. Setup time is the time that the signal has to be stable before the clock signal arrives. The state of the signal can be either low or high, but it has to be held at that level for the entire setup time. Only if this condition is met will the flip-flop or any logic gate be able to recognize the logic level. This is usually a very small time interval (of the order of $8-10*10^{-9}$ seconds), but it has to be maintained. The second is the hold time, which is the time after the clock signal arrives at the input. This time interval, along with the setup time interval, also has to be satisfied. This is also a very small time interval, but it also has to be maintained. The setup and the hold times (and their relation to the clock signal for a flip-flop) are shown in Figure 7.8. The setup time and the hold time do not have to be equal to each other. Setup and hold times for rising-edge signal may be different than those for a falling-edge signal.

These timing concepts are shown in Figure 7.8 for a rising clock signal. The input signal has to remain stable for the setup time before the clock signal arrives. The input must also stay stable for the hold time after the clock edge is received by the flip-flop. For the logic circuit to operate as we expect it to, the timing

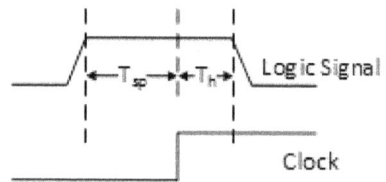

FIGURE 7.8. Setup and Hold times in a flip-flop.

of the setup time and the hold time must not be violated. If we violate the setup or the hold time, the circuit may recognize the input as we intended it to be. This is not our intent. We want the circuit to recognize the input that we applied, and not something else. In other words, you do not know what the circuit will recognize as the input if the setup and hold times are not satisfied. Setup and hold times have to be satisfied for all logic circuits, not just for flip-flops.

Review Question for Section 7.2

Question: Examine the timing diagram for an R S flip-flop shown in Figure 7.9. Draw the Q output for the flip-flop, assuming that the flip-flops are triggered as shown in the diagram.

Answer: The diagram also shows the expected output waveform for both the rising-edge triggered flip-flop and for the falling-edge triggered flip-flop. Notice that for the rising-edge triggered flip-flop, there is one clock period when the output from the flip-flop is unknown, as the input during that time interval is $(R\ S) \rightarrow (1\ 1)_2$ (which is not an allowed input). From the truth table and our study earlier, we know that during this time interval both the Q and the Q outputs will be low. Also note that the outputs from the two different triggering types are different, so it is very important to know what type of triggering is used for the flip-flop that you are using.

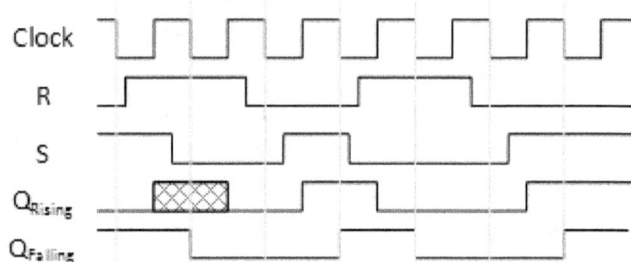

FIGURE 7.9. Output from flip-flops triggered differently.

7.3. OTHER TYPES OF FLIP-FLOPS

In the previous section we saw that the R S flip-flop has one combination of input that is not allowed. How can w be sure that that particular input will never occur in a design? We cannot guarantee that the R S flip-flop will never have $(1\ 1)_2$ as its input without changing the design of the flip-flop. In this section, we examine some other types of flip-flops and see how they are related to the R S flip-flop. To understand the operation of the flip-flops, we also need to have some notation that deals with time. This is due to the effect of the clock. The flip-flop responds only after the trigger event has occurred on the clock. To show the effects before and after the clock trigger, we will use the subscript n, so the quantity Q_n can be thought of as the present output, and the quantity Q_{n+1} can be thought of as the same signal or quantity, but after the trigger event on the clock. With this understanding, we will examine Figure 7.10.

S_n	R_n	Q_n	Q_{n+1}	
0	0	0	0	Hold
0	0	1	1	
0	1	0	0	Reset
0	1	1	0	
1	0	0	1	Set
1	0	1	1	
1	1	0	x	Not
1	1	1	x	Allowed

Truth Table for
S R Flip-Flop

J_n	K_n	Q_n	Q_{n+1}	
0	0	0	0	Hold
0	0	1	1	
0	1	0	0	Reset
0	1	1	0	
1	0	0	1	Set
1	0	1	1	
1	1	0	1	Toggle
1	1	1	0	

Truth Table for
J K Flip-Flop

SR

Q_n	00	01	11	10
0	0	0	x	1
1	1	0	x	1

$$Q_{n+1} = S + \overline{R}\,Q$$

JK

Q_n	00	01	11	10
0	0	0	1	1
1	1	0	0	1

$$Q_{n+1} = J\,\overline{Q} + \overline{K}\,Q$$

FIGURE 7.10. Truth Table and Logic Equation for S R and the J K flip-flop.

7.3.1. THE R S FLIP-FLOP

The truth table for the R S flip-flop has to include the current state of the flip-flop: since it has memory, it has to remember what the current state of the flip-flop is. To distinguish the current output from the next output from the flip-flop, we have used the subscript (n) for the current output and the subscript (n + 1) for the next output. From the truth table we have built the K-Map, which is also given in Figure 7.10. From the K-Map we can write the equation for the next state in terms of the output of the current state, as shown in Equation (7.1).

$$Q_{n+1} = S + \bar{R} \bullet Q_n \tag{7.1}$$

Equation (7.1) is a very compact way of describing the behavior of the S-R flip-flop. It relates the next output in terms of the present inputs and outputs. For example, the next output is going to be 1 when the S input is 1, independent of the other two inputs.

Note that when we say that the input $(1\ 1)_2$ is not allowed, it is like saying that this input will never occur. Since the input is never going to occur, it is treated as a "don't care" condition in writing the logic equation.

The J K Latch: As we mentioned earlier, engineers have come up with different and clever ways to make sure that we never have a $(1\ 1)_2$ input to the S R latch. One of these methods leads to the J K latch. The truth table and the logic equation of the J K latch are given in Figure 7.10. The first three conditions in the S R and the J K latch are the same. The J K latch has a memory condition when the inputs are $(0\ 0)_2$. The next condition is the reset condition: the reset of the latch occurs when the J K input is $(0\ 1)_2$. The third condition is the set condition: the latch is set when the J K input is $(1\ 0)_2$. The last condition, when the inputs are both $(1\ 1)_2$, was not allowed in the S R latch. In the J K latch, this input is allowed, and it toggles the output. If the previous output was a 0, then the next output will be a 1. Similarly, if the previous output was a 1, then the next output will be a 0. With the toggle condition, we have removed the restriction that the $(1\ 1)_2$ input is not allowed. With this change, the logic equation for the J K latch is given in Equation (7.2)

$$Q_{n+1} = J\bar{Q}_n + \bar{K}Q_n \tag{7.2}$$

It is interesting to compare the logic equation of the S R latch with the logic equation of the J K latch. To do this comparison, look at Figure 7.11. In Figure 7.11, we have used the R S latch to build a J K latch. From Figure 7.11, we see that the S input is $J\bar{Q}$, and the R input is KQ. Entering this information in the equation for the S R latch, we get Equation 7.3.

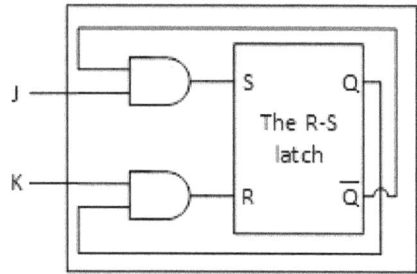

FIGURE 7.11. Using the S R latch to build a J K latch.

$$S = J \bullet \bar{Q}_n \qquad R = K \bullet Q_n$$

$$Q_{n+1} = S + \bar{R}Q_n = J \bullet \bar{Q}_n + \overline{\left(K \bullet Q_n\right)} \bullet Q_n = J \bullet \bar{Q}_n + \left(\bar{K} \bullet Q_n + \bar{Q}_n Q_n\right) = J \bullet \bar{Q}_n + \bar{K} \bullet Q_n$$

$$(7.3)$$

Equation (7.3) is the equation of the J K latch (which we obtained by substituting the logic functions for the S and the R inputs to the S R latch), so this gives us one neat way to use the S R latch while making sure that the S and the R inputs will never be $(1\ 1)_2$. This circuit, however, has its own problem. This is explained as follows.

You will remember from earlier how we explained that the R S latch became unstable and started to oscillate. That is, the output went from high to low, and then back to high, and then to low. It kept on doing this. The J K latch built using the R S latch has a similar (but not the same) problem. When the J K latch is placed in the toggle mode, then both the J and the K inputs are logic high. In this mode, the output toggles forever (or until we change one of the inputs). This happens because the latch is in the toggle mode: it is supposed to toggle. We can correct this in several different ways. To correct these oscillations of the latch, we can add a clock to the latch. The clock would be a third input to the AND gate. The AND gates provide the input to the S R latch, so the latch is converted to a flip-flop. With the clock present, the output of the AND gate is low when the clock is low. Also with the clock, the R S flip-flop will become a rising-edge triggered flip-flop. With the clock present, the R S flip-flop will see the input only at the rising edge of the clock. Since the flip-flop does not see the input when the clock is low, the output will not oscillate. Another way of correcting this

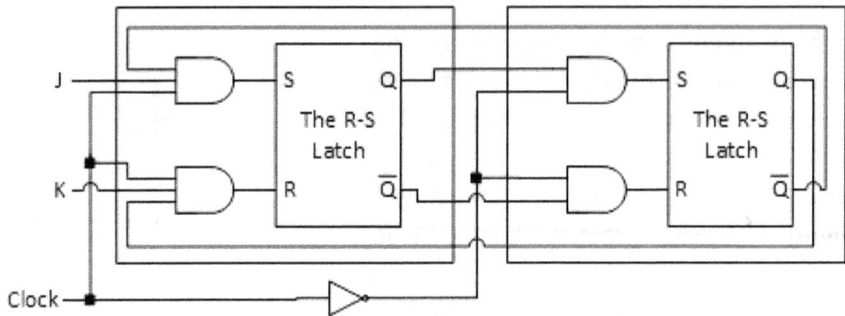

FIGURE 7.12. Using the S R latch to build a J K flip-flop.

oscillatory behavior is by the use of the Master–Slave latch. This is done in the exercises.

Question: How would you make the J K latch behave like a J K flip-flop, while still using the R S latch and not an R S rising-edge triggered flip flop?

>**Answer:** Figure 7.12 shows how to use an R S latch to overcome the oscillations when the J K latch is placed in the toggle mode. Now the output of the slave latch is fed back to the master latch. Since the slave output does not change until the second phase of the clock, the output from the master latch stays stable. When the first half of the clock is over, the slave responds and the toggle occurs only once per clock cycle.

The D Flip-flop: The J K flip-flop solves the problem that the R S flip-flop has when the two inputs are $(1\ 1)_2$ by making sure that the R and the S inputs never see $(1\ 1)_2$ as an input, even when the external input is $(1\ 1)_2$. Another way to avoid both the R and the S inputs being high at the same time is to connect the S and the R inputs to each other (but with an inverter in between them) and have only one input to the latch and the flip-flop. This is shown in Figure 7.13. For this flip-flop, with the inverter present, the R input is the complement of the S input, so the S and the R inputs are never $(1\ 1)_2$; hence, the not-allowed input never occurs. When the S input is D and the R input is $\bar{D}$, the input to the S R latch will either be $(0\ 1)_2$ or $(1\ 0)_2$. The logic equation for the D flip-flop can be obtained as shown in Equation (7.4)

$$S = D \qquad R = \bar{D}$$

$$Q_{n+1} = S + \bar{R}Q = D + \overline{(\bar{D})} \bullet Q = D \bullet (1 + Q) = D \tag{7.4}$$

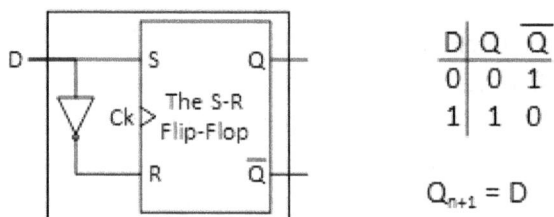

The logic diagram and truth table of the D Flip-Flop:

D	Q	$\overline{Q}$
0	0	1
1	1	0

$$Q_{n+1} = D$$

FIGURE 7.13. The logic diagram of the D Flip-Flop.

Just as the S R flip-flop can be either rising-edge triggered, falling-edge triggered, or Master–Slave, so also can the D flip-flop. The "D" in the D flip-flop stands for delay, so this flip-flop is also known as the delay flip-flop. The delay in this respect is explained as follows: the output follows the input after the clock delay. That is, the output will be the same as the input after the clock trigger has been received; hence, the output follows the input, but only after the appropriate delay.

The T Flip-flop: Another way to solve the problem of the R and S inputs both being logic high is to take the J K flip-flop and connect both the J and the K inputs together. This way, the inputs to the J K flip-flop will be either $(0\ 0)_2$, which is the memory condition, or they will be $(1\ 1)_2$, which is the toggle condition. The "T" in the T flip-flop stands for toggle flip-flop. The toggle in this respect is explained as follows: when the input to the flip-flop is logic high, the output changes state every clock period, or the output toggles. The logic diagram and the truth table of the T flip-flop are shown in Figure 7.14. The logic equation for the T flip-flop can be obtained as shown in Equation (7.5)

$$J = T \qquad K = T$$
$$Q_{n+1} = J\overline{Q} + \overline{K}Q = T\overline{Q} + \overline{T}Q \qquad (7.5)$$

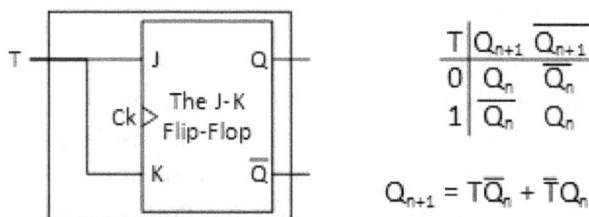

T	Q_{n+1}	$\overline{Q_{n+1}}$
0	Q_n	$\overline{Q_n}$
1	$\overline{Q_n}$	Q_n

$$Q_{n+1} = T\overline{Q_n} + \overline{T}Q_n$$

FIGURE 7.14. The logic diagram of the T Flip-Flop.

We have four different types of flip-flops which are all derived from the basic R S flip-flop. As we study the applications of the flip-flops, we will see that we can use any type of flip-flop to build the circuits that we want. The logic surrounding them will be different, but any of the flip-flops can be used to build the required logic function, since all the flip-flops have the same basic characteristic: all the flip-flops are able to set a logic high or low on the output as needed, and have a state that allows them to remember the last input.

In most of the designs that use flip-flops, the most frequently encountered flip-flops are the J K and the D flip-flops. This is because the combinational logic required to use either of these two flip-flops in designs is very frequently much less. It is true that the J K flip-flops require two inputs and the D flip-flops need only one input, but in determining the excitation required to change the state of the J K flip-flop, there are many "don't care" entries. (We will see this effect in the next chapter.) So when the requirement is to minimize the number of inputs, the D flip-flop is used; when the requirement is to minimize the gate count, the J K flip-flop is used. The T flip-flops are rarely available as packaged flip-flops, since they can easily be obtained from the J K flip-flops. The most obvious use of the T flip-flops is in building counters, which is the focus of the remaining portion of this chapter.

7.3.2. OTHER INPUTS ON THE FLIP-FLOPS

The 7474 is the IC that has two independent D flip-flops built into it. Each of the two flip-flops has its own clock inputs: the data input and the two outputs Q and $\overline{Q}$. In addition to these inputs, there are two more inputs. These are the $\overline{S_d}$ and the $\overline{R_d}$ inputs for each of the flip-flops. The two sets of inputs (D, clock, and the $\overline{S_d}$, $\overline{R_d}$) have some unique characteristics. The D input and the clock input are synchronous inputs. A synchronous input is recognized by the logic circuit only at a specific instant, like the rising edge or the falling edge of the clock signal. The $\overline{S_d}$, $\overline{R_d}$ inputs, on the other hand, are asynchronous inputs. An asynchronous input is recognized by the flip-flop as soon as it is applied. The flip-flop will respond to the asynchronous inputs any time they are presented to the flip-flop. The flip-flop does not wait for the synchronizing signal to respond to these inputs.

The function of these asynchronous inputs is to get the flip-flop in a known state when we first turn on the power to the circuit. When we first apply the power to the flip-flop, it can be in any state. This means that the output of the flip-flop may be logic high or it may be logic low. We do not know which state the flip-flop will be when we turn it on. The two asynchronous inputs enable us to place the flip-flop in the state we want to start up. The $\overline{S_d}$ input is used to set the flip-flop, so the Q output from the flip-flop will be logic high when the $\overline{S_d}$ input is activated by placing a logic low on this input. The $\overline{R_d}$ input is used to reset the flip-flop, so the Q output from the flip-flop will be logic low when the $\overline{R_d}$ input is activated by placing a logic low on this input. You must not activate both the $\overline{S_d}$ and $\overline{R_d}$ inputs at the same time. Also, these inputs must be cleared once they have placed the flip-flop in the state that you want it to be in. The asynchronous inputs are present in almost all the flip-flops. Sometimes they have a different name (for example, in the 7476 IC, which is a J K flip-flop, the two asynchronous inputs are labeled as $\overline{Pr}$, which stands for preset, and $\overline{Cr}$, which stands for clear): their function is to either set the flip-flop or to reset it.

Question: Draw the logic diagram, write the truth table, and write the logic equation for the four different types of flip-flops.

> **Answer:** The logic diagram, the truth table, and the logic equation for the R S flip-flop are shown in Figure 7.6. The logic diagram, the truth table, and the logic equation for the J K flip-flop are shown in Figure 7.10 and Figure 7.11. The logic diagram, the truth table, and the logic equation for the D flip-flop are shown in Figure 7.13. The logic diagram, the truth table, and the logic equation for the T flip-flop are shown in Figure 7.14.

7.4. USING FLIP-FLOPS AS REGISTERS

The use of flip-flops in sequential circuits is the most common use of the flip-flops; this we will see in the next chapter. Another use of flip-flops is as registers. A register is a group of storage elements which are read or written as a unit. The simplest way to do this is to use a common clock for all the flip-flops that make up a register. The following two different applications show some of the uses of the flip-flops.

7.4.1. USING FLIP-FLOPS AS A STORAGE REGISTER

The most basic flip-flops are also available in IC's as octal flip-flops. This means that there are eight flip-flops in this IC. They all have a common clock input, along with a common preset and clear input. This means that all eight flip-flops are triggered simultaneously, and all are set to some known state that is the same for all the flip-flops. Such an IC is referred to as an *Octal Register*. An example of this is the 74LS273. This IC contains eight D flip-flops, all of which are triggered on the positive edge, and it has an active low Master reset so the flip-flops can only be cleared asynchronously. One application of the 74LS273 is to hold data to drive a seven-segment display. This is shown in Figure 7.15.

Figure 7.15 shows an octal register of D flip-flops. The input for this register most probably comes from a BCD to seven-segment decoder like the one we saw in Chapter 6. The Master reset is tied to logic high so that input is deactivated. When the clock pulse arrives, the new data will be placed into the eight D flip-flops. The output of the flip-flops will then light up the correct segments to display the desired digit.

7.4.2. USING FLIP-FLOPS AS A SHIFT REGISTER

A register can be used to change the format of data, from serial to parallel or from parallel to serial. In the application where we want to change the format from serial to parallel, data arrives at the input of the register one bit at a time. The register shifts the data in at one end of the register (as shown in Figure 7.16). When all the bits for one data item are gathered in the register, they are presented to the other part of the circuit all together or in parallel. These registers that

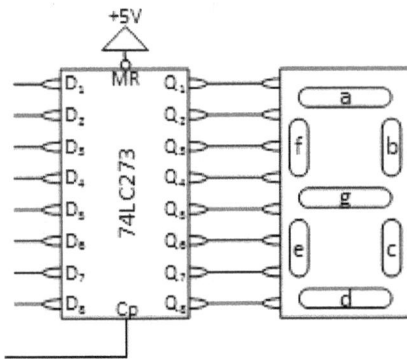

FIGURE 7.15. Using a register to hold display data.

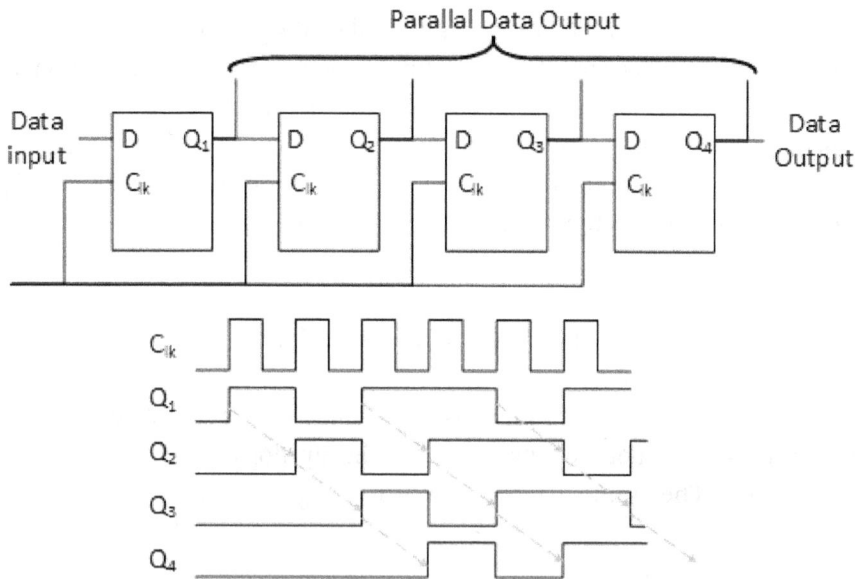

FIGURE 7.16. Data shifting in a shift register.

shift and store bits of information are called *Shift Registers*. A typical four-bit shift register is shown in Figure 7.16. A shift register can also be used to take a group of data bits and shift them out, one bit at a time, to some other part of the circuit.

The flip-flops in Figure 7.16 are assumed to begin with logic low output. This would happen by the use of the asynchronous clear input. Next, with each clock pulse, data arrives at the input of flip-flop Q_1. In the figure shown, the arriving data is $(1\ 0\ 1\ 1\ 0\ 1)_2$, so after one clock pulse, the four flip-flops will be $(Q_1\ Q_2\ Q_3\ Q_4) \rightarrow (1\ 0\ 0\ 0)_2$, and we see that the first data bit has been shifted into the shift register. After the second clock pulse, the four flip-flops will be $(Q_1\ Q_2\ Q_3\ Q_4) \rightarrow (0\ 1\ 0\ 0)_2$, and we see that the first two data bits have been shifted into the shift register. After the third clock pulse, the four flip-flops will be $(Q_1\ Q_2\ Q_3\ Q_4) \rightarrow (1\ 0\ 1\ 0)_2$, and we see that the first three data bits have been shifted into the shift register. After the fourth clock pulse, the four flip-flops will be $(Q_1\ Q_2\ Q_3\ Q_4) \rightarrow (1\ 1\ 0\ 1)_2$, and we see that the first four data bits have been shifted into the shift register. This continues as long as new data is arriving. In this register, we also have the ability to read all the data that is in the shift register. This is shown as *Parallel Data Output*.

The shift register in Figure 7.16 is a serial load right shift register. Other shift registers can have parallel load, shift left or shift right, or any combination. Shift registers are typically used in modems where the microprocessor has the data available as several bits but the modem can only transmit one bit at a time, or where the modem can receive data one bit at a time but the microprocessor needs to use all the bits simultaneously.

Review Question for Section 7.4

Question: For an eight-bit shift register, the input data is $(01100101)_2$. Draw the timing diagram of the output from all the flip-flops.

Answer: The timing diagram is shown in Figure 7.17.

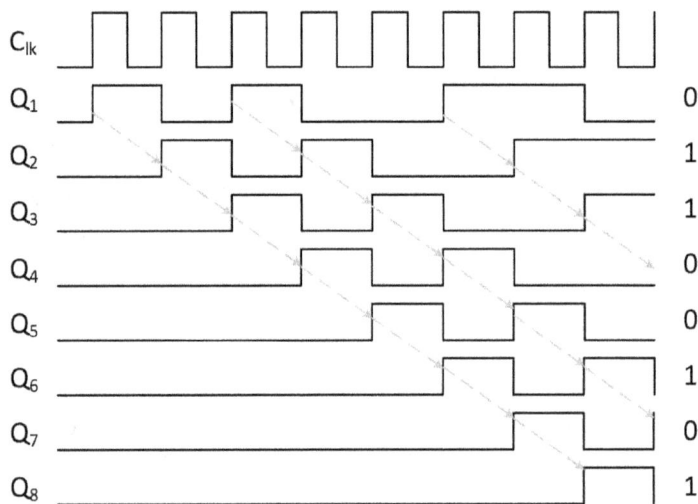

FIGURE 7.17. Data shifting through an eight bit shift register.

7.5. CHAPTER PROBLEMS

7.7.1. Build a feedback circuit with cross-coupled NAND gates. Develop a truth table for this circuit. From the truth table, identify: the

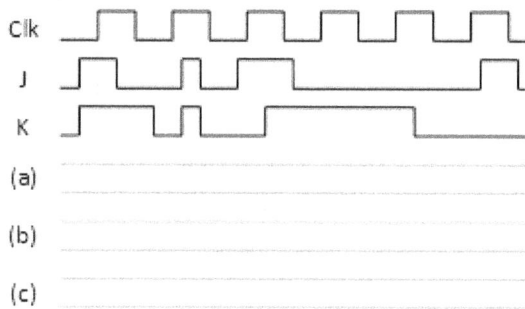

FIGURE 7.18. J K flip-flop input waveform.

memory input; set input condition; reset input condition; and the not-allowed input condition.

7.7.2.   Show how you would modify a J K flip-flop to use as a D flip-flop. Show this using the logic equation, and then show the logic diagram.

7.7.3.   You are given the clock waveform and the J K input in Figure 7.18. Draw the output waveform when the J K flip-flop is: (a) positive edge triggered flip-flop; (b) negative edge triggered flip-flop; (c) Master–Slave flip-flop. Assume the flip-flop is reset initially.

7.7.4.   You are given the clock waveform and the D input in Figure 7.19. Draw the output waveform when the D flip-flop is: (a) positive edge triggered flip-flop; (b) negative edge triggered flip-flop; and (c) Master—Slave flip-flop. Assume the flip-flop is reset initially.

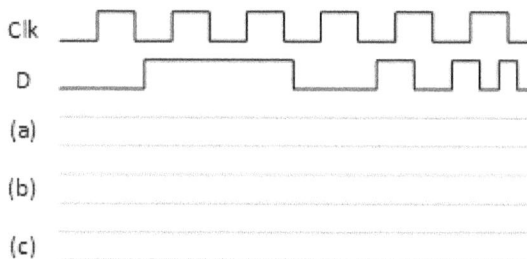

FIGURE 7.19. D flip-flop input waveform.

7.7.5. Is it true that any flip-flop type can be used as any other flip-flop type with proper logic gates used for conversion? Show how you would convert the J K flip-flop to all the other types of flip-flops.

7.7.6. Very often, a PAL or PLA device has a flip-flop after the OR gate and before the output. This flip-flop is usually a D flip-flop. Draw the diagram of a PAL device with the flip-flop.

7.7.7. Very often in digital systems, there is the need for two different clocks that operate at the same frequency but are 180 degrees out of phase. Show how you would produce two clock signals which are 180 degrees out of phase with each other by using cross-coupled NOR gates.

7.7.8. In many sequential circuits, there is the need to change the output of a flip-flop from one specific output to a different specific output. For example, if the present output is 0 and we want the next output to be a 1, then we want the flip-flop to change its output from $0 \rightarrow 1$. To achieve this, we must have a specific input based on the type of flip-flop we have. For this problem, determine all the possible inputs that you can use for this transition. Do this for all the four different types of flip-flops.

7.7.9. Repeat Question 7.7.8 for a transition from 1 to 0.

7.7.10. Repeat Question 7.7.8 for a transition from 1 to 1.

7.7.11. Repeat Question 7.7.8 for a transition from 0 to 0.

8. COUNTERS AND SEQUENTIAL LOGIC

8.0. INTRODUCTION

Until now, we have seen logic circuits whose outputs are solely the function of the present inputs. Circuits whose outputs depend only on the present input are known as combinational circuits. We can also have logic circuits whose output depends on the present and all the past inputs. These circuits must store the past inputs in some form of history. The information about the past inputs is stored in the state of the machine. Circuits whose outputs depend on the past and the present inputs are known as sequential circuits. To store the state of the machine, the circuits use memory elements; these memory elements, the flip-flops, were the subject of the previous chapter. In this chapter, we extend our understanding of the flip-flop and its use in sequential circuits.

Sequential execution of events is a very common occurrence in our lives. We are always timing things or setting the order in which we will execute various tasks that we have to complete. The traffic light at the street corner is a sequential circuit. In a computer system, a sequential machine executes tasks sequentially by setting priorities or issuing timing signals. In this chapter, we will first see how to build a simple sequential machine whose input is only a clock signal. We call this sequential machine a counter. Then we will investigate a sequential machine that has external inputs and whose output depends on the current state of the machine and the input that the machine receives.

8.1. DESIGNING COUNTERS

A counter is a circuit that keeps track of the number of events that have occurred. Every time an event occurs, a clock pulse is provided to the circuit and the circuit goes to the next state, thus remembering or counting the total number of events that have taken place.

8.1.1. A SIMPLE EXAMPLE OF A COUNTER

A very simple example of a counter is shown in Figure 8.1. Assume that the outputs of all three flip-flops are zeros to begin with. The J K inputs for flip-flop Q_3 are taken from the output of flip-flop Q_1, $Q_1 \rightarrow K_3$, and $\overline{Q}_1 \rightarrow J_3$.

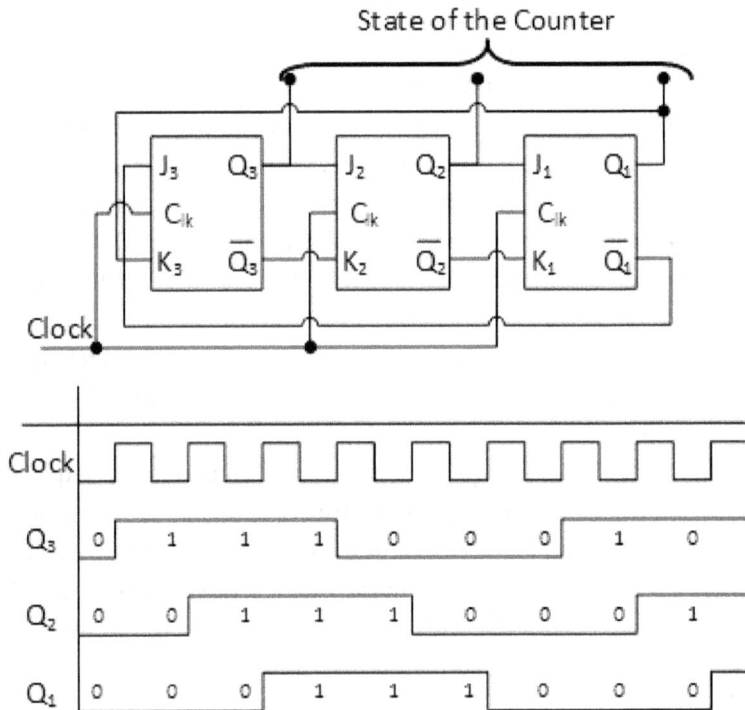

FIGURE 8.1. A three-bit Johnson Counter.

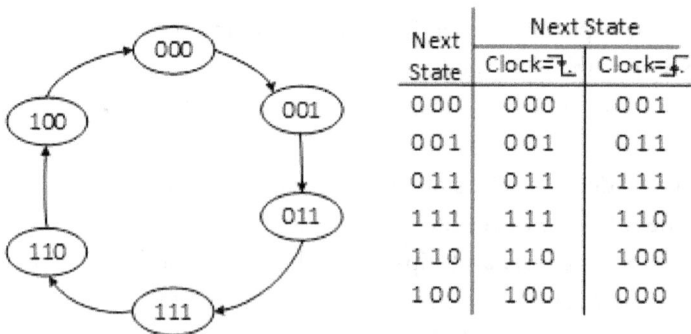

Next State	Next State Clock=↑	Clock=↓
0 0 0	0 0 0	0 0 1
0 0 1	0 0 1	0 1 1
0 1 1	0 1 1	1 1 1
1 1 1	1 1 1	1 1 0
1 1 0	1 1 0	1 0 0
1 0 0	1 0 0	0 0 0

FIGURE 8.2. The State Transition Diagram and the State Transition Table of a Johnson counter shown in Figure 8.1.

During the first clock pulse, this connects a high to J_3 and a low to K_3. This is a set condition for the J K flip-flop. After the next event, (clock) flip-flop Q_3 will be set. Flip-flop Q_2 follows the output that flip-flop Q_3 had before the current event, and flip-flop Q_1 follows the output flip-flop Q_2 had before the current event occurred. This happens after every clock pulse. With this, the output sequence of the counter can be represented as shown in Figure 8.2.

The diagram on the left side of Figure 8.2 is known as the *State Transition Diagram*, or the state diagram. It tells us what the output of the flip-flop will be after each event (in this example, after each clock pulse). The diagram tells us that the state machine (or our counter) will transition from state 000 to state 001 when the first event takes place, or when the first rising edge of the clock pulse arrives. When we are designing a counter or a sequential machine, this is generally the first step: we begin with a state transition diagram. This diagram gives us the big picture of what our machine is going to do. It is difficult to do much with this diagram, so the next step in building the machine is to draw a *State Transition Table*. The state transition table for our counter is drawn on the right side of Figure 8.2.

The state transition table tells us under what condition the machine will go to the next state; it also tells us what the next state is supposed to be. For our current example, when the clock input is a rising edge, there is a transition in the state machine. When the clock input is a falling edge, there is no transition in the machine. The next state that is listed is the state that the machine will transition to. There is not much difference between the state transition diagram and the state transition table: the information is the same, but it is packaged in a different way. Using the state transition table, we can determine what the flip-flop input should be to achieve the required transition. The required transition is determined for each flip-flop. We do this as follows.

We have identified the state of the machine as the output of the flip-flop. When the machine is in state 000, flip-flop Q_1 has its output zero, flip-flop Q_2 has its output zero, and flip-flop Q_3 has its output zero. When the machine is in state 000 and we receive a rising edge of the clock, the machine will transition to the state 001. This means that the flip-flop Q_3 has to change from state zero to state one. The output of the other two flip-flops has to remain the same. As another example, consider the present state 110. This time flip-flop Q_1 has its output one, flip-flop Q_2 has its output one, and flip-flop Q_3 has its output zero. When the machine transitions on the rising edge, the machine goes to state 100. This requires flip-flop Q_2 to transition from an output one to output zero. This transition in the output of the flip-flop takes place when there is a proper input to the flip-flop. The next question that we want to answer is: How do we get this proper input to the flip-flop in order to get the transition that we want?

To determine the correct input, we will build the excitation table. The excitation table tells us what we need to have (as the input to the flip-flop) for the output of the flip-flop to change from its current output to the next output that we want. To build the excitation table for the J K flip-flop that we used in the state machine in Figure 8.1, look at Figure 8.3. There are four possible combinations of the present output and the next output. These four combinations are listed in the excitation table under the columns for Q_n and Q_{n+1}. The J and the K inputs for all four transitions are determined in Figure 8.3.

As an example, consider the transition from present output $Q_n = 0$ to the next output $Q_{n+1} = 0$. We can accomplish this in one of two ways. First, when both the J and the K inputs are zero, then the flip-flop output does not change. This implies that Q_{n+1} will be zero when Q_n is also zero. Second, we can get a zero output for Q_{n+1} by making $J = 0$ and $K = 1$. This is the reset condition. Combining these two possibilities together, we can conclude that the J K inputs should be 0 for J and "don't care" for K. This is shown in Figure 8.3, Table a. The other three combinations are obtained in the same way. They are shown in Tables b, c, and d in Figure 8.3. With the excitation table for the flip-flop, we can determine the input for any specific transition.

J	K	Q_{n+1}
0	0	Q_n
0	1	0
1	0	1
1	1	$\overline{Q_n}$

Table a

For a 0 to 0 transition set J to zero and K to don't care.

J	K	Q_{n+1}
0	0	Q_n
0	1	0
1	0	1
1	1	$\overline{Q_n}$

Table b

For a 0 to 1 transition set J to one and K to don't care.

Q_n	Q_{n+1}	J	K
0	0	0	x
0	1	1	x
1	0	x	1
1	1	x	0

Excitation Table for J K flip-flop.

J	K	Q_{n+1}
0	0	Q_n
0	1	0
1	0	1
1	1	$\overline{Q_n}$

Table c

For a 1 to 0 transition set J to don't care and K to one.

J	K	Q_{n+1}
0	0	Q_n
0	1	0
1	0	1
1	1	$\overline{Q_n}$

Table d

For a 1 to 1 transition set J to don't care and K to zero.

FIGURE 8.3. Building the excitation table for the J K flip-flop.

The next step in building our counter is to use the state transition table to determine the excitation for each flip-flop. Once we have determined the required excitation, the machine can be built. We do this with the aid of Figure 8.4a.

In Figure 8.4a, we determine the input required for flip-flop Q_1. When the clock goes from low to high, the flip-flop knows that this is an event, and it will check its inputs and set the next output according to the inputs. To determine the correct input to flip-flop Q_1, we look at only what is happening to flip-flop Q_1. We see that when the counter is in state $(0\ 0\ 0)_2$ it will transition to state $(0\ 0\ 1)_2$. As far as flip-flop Q_1 is concerned, this represents a 0 à 0 transition. When there is a $0 \rightarrow 0$ transition on a J K flip-flop, we must connect a 0 to the J input and a "don't care: to the K input, according to the excitation table for the J K flip-flop in Figure 8.3. This is entered in the two tables being constructed in Figure 8.4a. We use these tables to build a logic equation for the J and the K inputs of flip-flop Q_1. Next, when the counter is in state $(0\ 0\ 1)_2$, it will transition to state $(0\ 1\ 1)_2$. This is also a $0 \rightarrow 0$ transition for flip-flop Q_1. When there is a $0 \rightarrow 0$ transition on a J K flip-flop, we must have a zero input to the J input and a "don't care" input to the K. Next, when the counter is in state $(0\ 1\ 1)_2$, it will transition to state $(1\ 1\ 1)_2$. This

Present State	Next State Clock=↑	Next State Clock=↓
000	000	001
001	001	011
011	011	111
111	111	110
110	110	100
100	100	000

Output of flip-flop Q_1

Present State	J input for flip-flop Q_1 Clock=↓
000	0
001	0
011	1
111	x
110	x
100	x

Present State	K input for flip-flop Q_1 Clock=↓
000	x
001	x
011	x
111	0
110	0
100	1

$Q_1 Q_2$ / Q_3

	00	01	11	10
0	0	x	x	x
1	0	1	x	x

J input for Q_1 $J_1 = Q2$

$Q_1 Q_2$ / Q_3

	00	01	11	10
0	x	x	0	1
1	x	x	0	x

K input for Q_1 $K_1 = \overline{Q_2}$

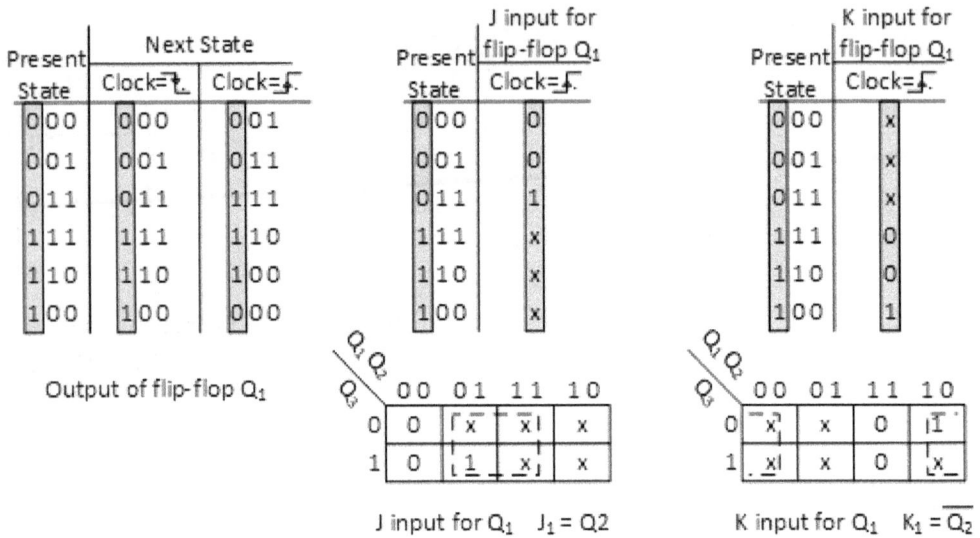

FIGURE 8.4A. Determining the input for flip-flop Q_1.

is a $0 \rightarrow 1$ transition. When there is a $0 \rightarrow 1$ transition on a J K flip-flop, we must have a one input to the J input and a "don't care" input to the K. This way, using the excitation table, the entire table for the required inputs to the J and the K inputs is filled out.

Once we have the table filled for all three flip-flops (as shown in Figures 8.4b and 8.4c), we need to obtain a logic equation for each of the inputs. To get this logic equation, we draw a K-map for each of the inputs. The K-maps for the J and K inputs for flip-flop Q_1 are drawn under the table in Figure 8.4a. Notice that, since the counter uses only six states in the machine, we will have excitation for only these six states. The two states that are not used will be "don't care" in the K-map for both the J and the K input. Each of the squares of the K-Map represents a state of the counter. Therefore, the value entered in each square of the K-map is the input that we wish to provide for the flip-flop when the counter is in that state.

The tables in Figure 8.4b give you the required information for flip-flop Q_2, and the tables in Figure 8.4c give you the required information for flip-flop Q_3. Comparing the logic equations for the three flip-flops with the diagram of the counter, we see how the counter was designed.

J input for flip-flop Q₂ — tables.

Figure 8.4B

Present State	Next State Clock=↑	Next State Clock=↓
000	000	001
001	001	011
011	011	111
111	111	110
110	110	100
100	100	000

Output of flip-flop Q_2

Present State	J input for flip-flop Q_2 Clock=↓
000	0
001	0
011	1
111	x
110	x
100	x

Present State	K input for flip-flop Q_2 Clock=↓
000	x
001	x
011	x
111	0
110	0
100	1

J input K-map (Q_1Q_2 across, Q_3 down):

Q_3 \ Q_1Q_2	00	01	11	10
0	0	x	x	0
1	1	x	x	x

J input for Q_2 $J_2 = Q_3$

K input K-map:

Q_3 \ Q_1Q_2	00	01	11	10
0	x	x	1	1
1	x	0	0	x

K input for Q_2 $K_2 = \overline{Q_3}$

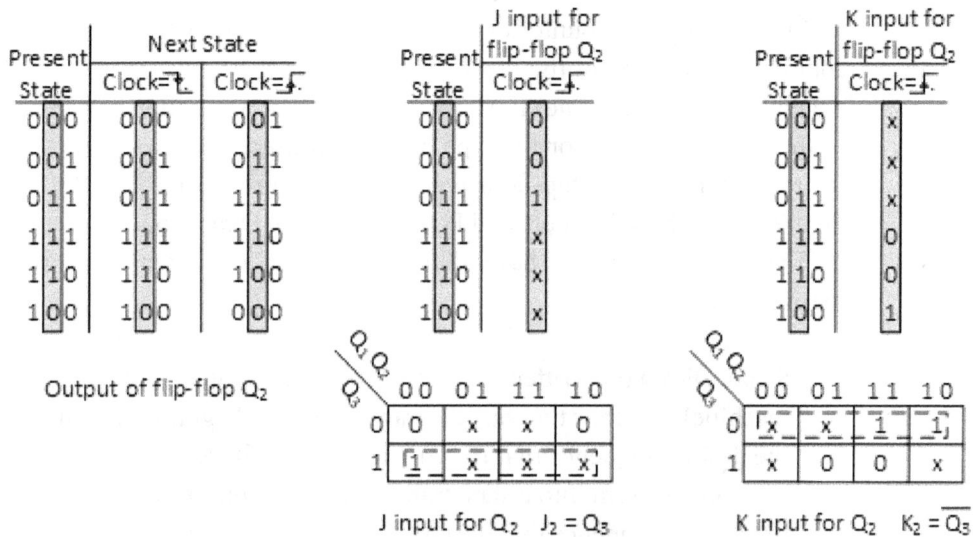

FIGURE 8.4B. Determining the input for flip-flop Q_2.

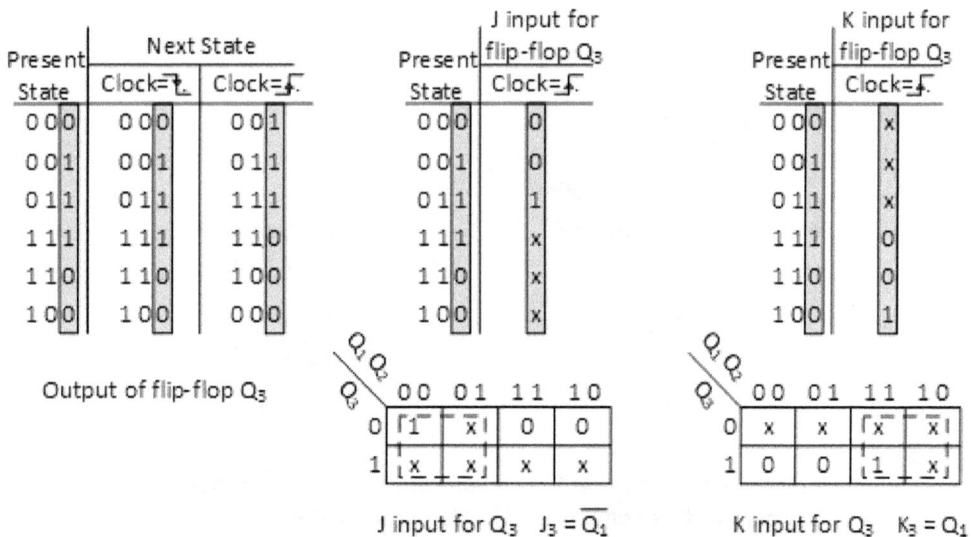

Figure 8.4C

Present State	Next State Clock=↑	Next State Clock=↓
000	000	001
001	001	011
011	011	111
111	111	110
110	110	100
100	100	000

Output of flip-flop Q_3

Present State	J input for flip-flop Q_3 Clock=↓
000	0
001	0
011	1
111	x
110	x
100	x

Present State	K input for flip-flop Q_3 Clock=↓
000	x
001	x
011	x
111	0
110	0
100	1

J input K-map (Q_1Q_2 across, Q_3 down):

Q_3 \ Q_1Q_2	00	01	11	10
0	1	x	0	0
1	x	x	x	x

J input for Q_3 $J_3 = \overline{Q_1}$

K input K-map:

Q_3 \ Q_1Q_2	00	01	11	10
0	x	x	x	x
1	0	0	1	x

K input for Q_3 $K_3 = Q_1$

FIGURE 8.4C. Determining the input for flip-flop Q_3.

In designing the counter of Figure 8.1, we used the J K flip-flops. We could have used any of the other flip-flops as well. The design procedure would have been the same. The only difference that we would see is the use of the excitation table for the type of flip-flop used. The excitation tables for the other types of flip-flops are given in Figure 8.5 for R S flip-flop, Figure 8.6 for the T flip-flop, and Figure 8.7 for the D flip-flop.

The procedure that we have followed to design a counter can be summarized as follows:

Step 1. From the written specifications of the counter (or the state machine), we draw the state diagram. The state diagram shows us the desired sequence that the counter has to follow. We translate the state diagram into a state transition table. Both of these show the desired sequence that the counter is expected to follow.

Step 2. The count sequence of the states tells us the output of each of the flip-flops. Using this information and the excitation tables of the flip-flops, we determine the input logic equation that is required (for each flip-flop that the counter will be using) by

S	R	Q_{n+1}
0	0	Q_n
0	1	0
1	0	1
1	1	--

For a 0 to 0 transition set S to zero and R to don't care.

S	R	Q_{n+1}
0	0	Q_n
0	1	0
1	0	1 ←
1	1	--

For a 0 to 1 transition set S to one and R to zero.

S	R	Q_{n+1}
0	0	Q_n
0	1	0 ←
1	0	1
1	1	--

For a 1 to 0 transition set S to zero and R to one.

S	R	Q_{n+1}
0	0	Q_n
0	1	0
1	0	1
1	1	--

For a 1 to 1 transition set S to don't care and R to zero.

Q_n	Q_{n+1}	S	R
0	0	0	x
0	1	1	0
1	0	0	1
1	1	x	0

Excitation Table for S R flip-flop.

FIGURE 8.5. Building the excitation table for the S R flip-flop.

T	Q_{n+1}
0	Q_n ←
1	$\overline{Q}_n$

For a 0 to 0 transition set T to zero.

T	Q_{n+1}
0	Q_n ←
1	$\overline{Q}_n$

For a 0 to 1 transition set T to one.

Q_n	Q_{n+1}	T
0	0	0
0	1	1
1	0	1
1	1	0

Excitation Table for T flip-flop.

T	Q_{n+1}
0	Q_n ←
1	$\overline{Q}_n$

For a 1 to 0 transition set T to one.

T	Q_{n+1}
0	Q_n ←
1	$\overline{Q}_n$

For a 1 to 1 transition set T to zero.

FIGURE 8.6. Building the excitation table for the T fli.

D	Q_{n+1}
0	0 ←
1	1

For a 0 to 0 transition set D to zero.

D	Q_{n+1}
0	0
1	1 ←

For a 0 to 1 transition set D to one.

Q_n	Q_{n+1}	D
0	0	0
0	1	1
1	0	0
1	1	1

Excitation Table for D flip-flop.

D	Q_{n+1}
0	0 ←
1	1

For a 1 to 0 transition set D to one.

D	Q_{n+1}
0	0
1	1 ←

For a 1 to 1 transition set D to zero.

FIGURE 8.7. Building the excitation table for the D flip-flop.

using a K-Map. This gives us a combinational logic function for the inputs of the flip-flops.

Setp 3. Using the combinational logic function obtained in Step 2, we build the counter that we want, using the flip-flops as memory elements.

Review Questions for Section 8.1.1

Question: Using the state transition diagram in Figure 8.8, build the state transition table.

Answer: The State transition table is shown in Figure 8.8.

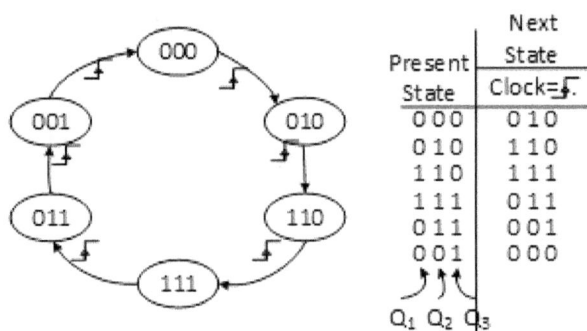

Present State	Next State Clock=↧
0 0 0	0 1 0
0 1 0	1 1 0
1 1 0	1 1 1
1 1 1	0 1 1
0 1 1	0 0 1
0 0 1	0 0 0

Q_1 Q_2 Q_3

FIGURE 8.8. The State Transition Diagram and the State Transition Table of a 3-bit counter.

Question: Draw the excitation tables for all three flip-flops, assuming that we will use the D flip-flop for all three flip-flops.

Answer: The excitation tables for all three flip-flops are shown in Figure 8.9.

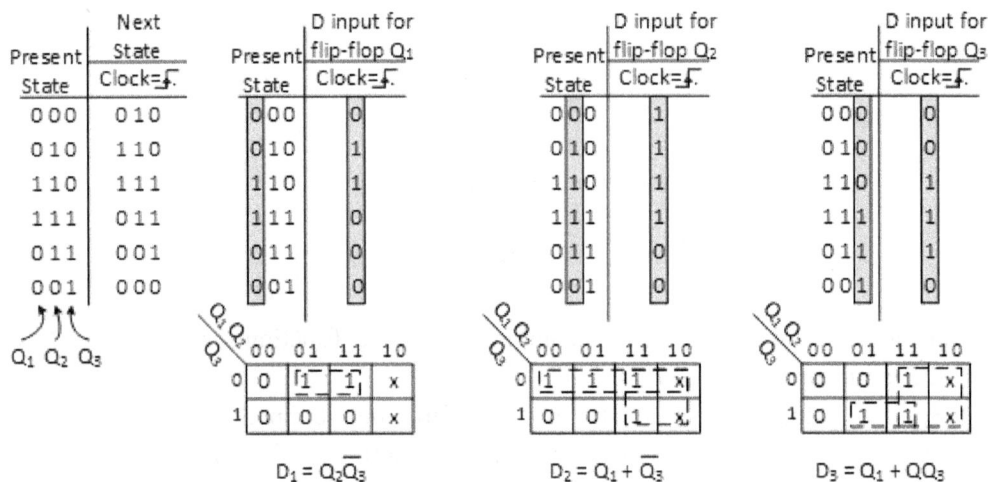

Present State	Next State Clock=↧
0 0 0	0 1 0
0 1 0	1 1 0
1 1 0	1 1 1
1 1 1	0 1 1
0 1 1	0 0 1
0 0 1	0 0 0

Q_1 Q_2 Q_3

Present State	D input for flip-flop Q_1 Clock=↧
0 0 0	0
0 1 0	1
1 1 0	1
1 1 1	0
0 1 1	0
0 0 1	0

Q_1Q_2 / Q_3	00	01	11	10
0	0	1	1	x
1	0	0	0	x

$D_1 = Q_2\overline{Q_3}$

Present State	D input for flip-flop Q_2 Clock=↧
0 0 0	1
0 1 0	1
1 1 0	1
1 1 1	1
0 1 1	0
0 0 1	0

Q_1Q_2 / Q_3	00	01	11	10
0	1	1	1	x
1	0	0	1	x

$D_2 = Q_1 + \overline{Q_3}$

Present State	D input for flip-flop Q_3 Clock=↧
0 0 0	0
0 1 0	0
1 1 0	1
1 1 1	1
0 1 1	1
0 0 1	0

Q_1Q_2 / Q_3	00	01	11	10
0	0	0	1	x
1	0	1	1	x

$D_3 = Q_1 + QQ_3$

FIGURE 8.9. Determining the input for three D flip-flops of a 3-bit counter.

8.1.2. COUNTERS WITH COMPLEX SEQUENCES

To see how the process described in Section 8.1.1 is used, we will look at another implementation of a counter. For this implementation, we will build a counter that follows the count sequence 000 → 101 → 010 → 110 → 011 → 000. This is a five-state counter. In this counter, we have three combinations (001; 100; and 111) that are not part of the count sequence. According to our design procedure, Step 1 is to draw the state diagram, then draw the state transition table from the state transition diagram. This is done in Figure 8.10. Step 2 requires us to use the state transition table, along with the excitation table for the flip-flops chosen, to build the input equation for the flip-flops. At this point we can choose the type of flip-flop that we want to use to build our counter.

In all the examples until now, we have used the same type of flip-flop to build the counter. This is not necessary; we can choose different flip-flops if we so desire. Most of the time, however, we will choose the same type of flip-flop to build our state machine. This, however, is not necessary. For our design, we need three flip-flops: this time we will choose the T flip-flops. Using the excitation table of the T flip-flop, we determine the excitation logic equation for the T input to the three flip-flops. This is done in Figure 8.11. Now that we have the logic equations for the input to the three flip-flops, we can build the counter using the T flip-flops and the logic functions. This is shown in Figure 8.12.

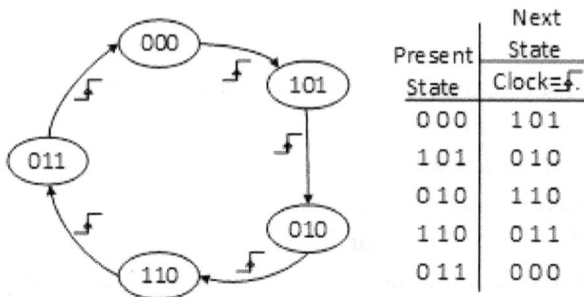

Present State	Next State Clock=⎍
0 0 0	1 0 1
1 0 1	0 1 0
0 1 0	1 1 0
1 1 0	0 1 1
0 1 1	0 0 0

FIGURE 8.10. The State Transition Diagram and the State Transition Table for a 5 state counter.

Present State	Next State Clock=↓
0 0 0	1 0 1
1 0 1	0 1 0
0 1 0	1 1 0
1 1 0	0 1 1
0 1 1	0 0 0

Q_1 Q_2 Q_3

T input for flip-flop Q_1

Present State	Clock=↓
0 00	1
1 01	1
0 10	1
1 10	1
0 11	0

Q_1Q_2 \ Q_3	00	01	11	10
0	1	1	1	x
1	x	0	x	1

$$T_1 = Q_1 + \overline{Q}_3$$

T input for flip-flop Q_2

Present State	Clock=↓
0 0 0	0
1 0 1	1
0 1 0	0
1 1 0	0
0 1 1	1

Q_1Q_2 \ Q_3	00	01	11	10
0	0	0	0	x
1	x	1	x	1

$$T_2 = Q_3$$

T input for flip-flop Q_3

Present State	Clock=↓
0 0 0	1
1 0 1	1
0 1 0	0
1 1 0	1
0 1 1	1

Q_1Q_2 \ Q_3	00	01	11	10
0	1	0	1	x
1	x	1	x	1

$$T_3 = Q_1 + \overline{Q}_2 + Q_3$$

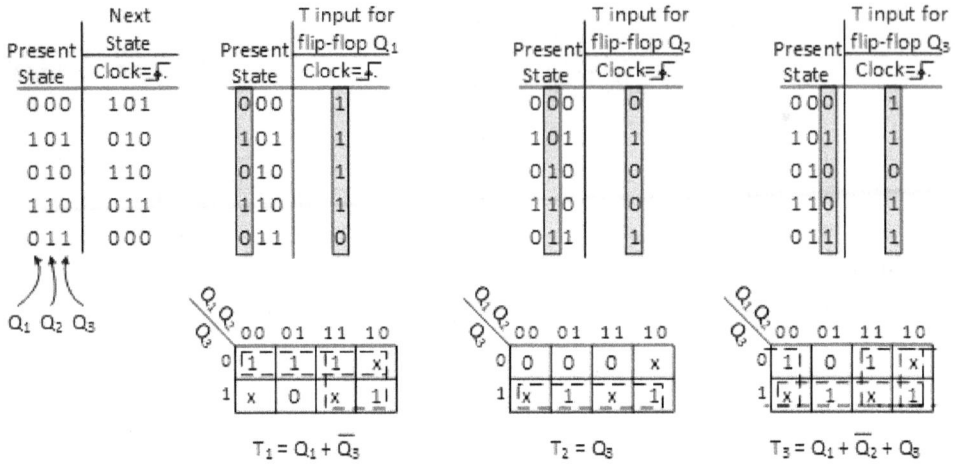

FIGURE 8.11. Determining the input for three T flip-flops of a 5 state counter.

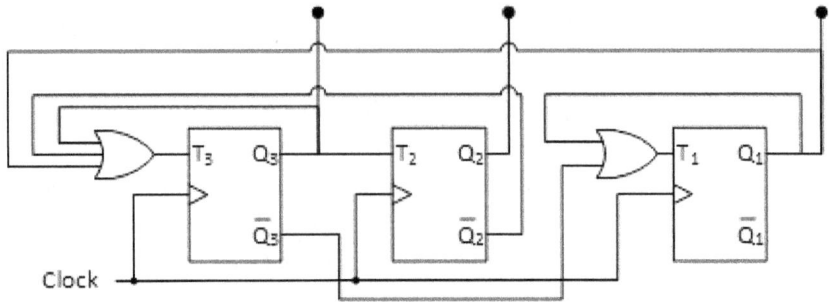

FIGURE 8.12. Logic diagram of a complex 5 state counter.

Figure 8.12 shows the logic diagram of the entire counter. A couple of things to consider for this counter are as follows. First, in what state does the counter start up? We know when power is applied to the flip-flops, the flip-flops come up as either set or cleared; we do not know which flip-flop will be set or which flip-flop will be clear. Therefore, it is possible that the counter could start up in one of the states that is not used. What will happen to the counter in this case? Will the counter be stuck in one of these unused states, or will it somehow transition to one of the counter states? If the counter gets stuck in one of the unused states, then the counter is not much use. To make sure that the counter starts up in a known state (and not stuck in one of the unused states), we can choose one of two options.

First, we can use the preset and the clear inputs that are present on the flip-flops to make sure that the flip-flops start out in a known state. To use this option, we must set the flip-flops in a known state when power is first applied to the entire circuit. Once the flip-flops are in the known state, the preset and clear inputs have to be changed to their passive states so the flip-flops will follow the counter sequence and will no longer be influenced by the preset and the clear inputs.

The second method is more elegant, and is the recommended method. To place the counter in the count sequence, we use the unused states as part of the counter so the counter can transition from these unused states to one of the count states. This way, the counter will never get stuck in one of the unused states. This concept is shown in Figure 8.13. Figure 8.13 shows what could happen if the counter starts out in one of the unused states. The idea is that if the counter starts in any one of the unused states, the counter will transition to one of the count states in just one clock cycle. The required modification is shown in the state transition diagram and the state transition table in Figure 8.13. With the change in the state transition table, we will have to reevaluate the excitation tables for all the flip-flops. This is shown in Figure 8.14. Notice that the process for designing the

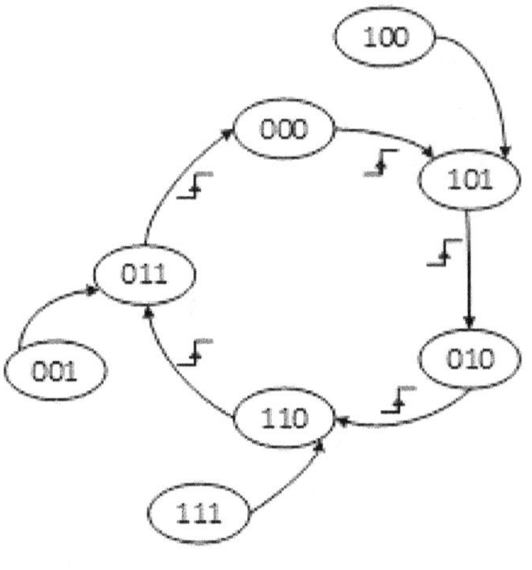

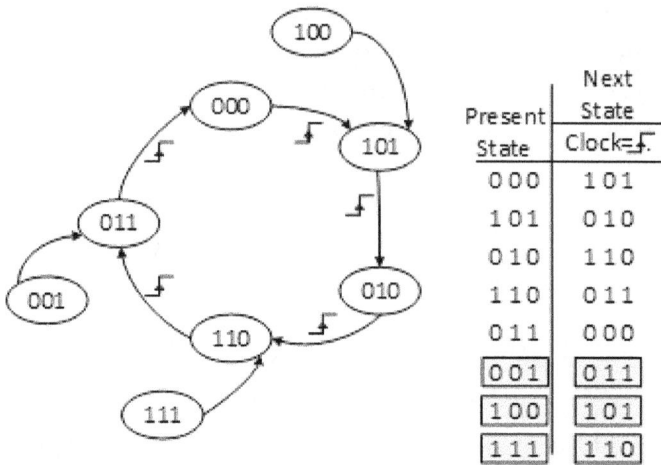

Present State	Next State Clock=↑
000	101
101	010
010	110
110	011
011	000
001	011
100	101
111	110

FIGURE 8.13. Designing a self starting counter.

Present State	Next State Clock=↓
000	101
101	010
010	110
110	011
011	000
001	011
100	101
111	110

$Q_1\ Q_2\ Q_3$

Present State	T input for flip-flop Q_1 Clock=↓
000	1
101	1
010	1
110	1
011	0
001	0
100	0
111	0

$Q_1Q_2 \backslash Q_3$	00	01	11	10
0	1	1	1	0
1	0	0	0	1

$$T_1 = Q_2\overline{Q_3} + Q_1\overline{Q_3} + Q_1\overline{Q_2}Q_3$$

Present State	T input for flip-flop Q_2 Clock=↓
000	0
101	1
010	0
110	0
011	1
001	1
100	0
111	0

$Q_1Q_2 \backslash Q_3$	00	01	11	10
0	0	0	0	0
1	1	1	0	1

$$T_2 = \overline{Q_1}Q_3 + \overline{Q_2}Q_3$$

Present State	T input for flip-flop Q_3 Clock=↓
000	1
101	1
010	0
110	1
011	1
001	0
100	1
111	1

$Q_1Q_2 \backslash Q_3$	00	01	11	10
0	1	0	1	1
1	0	1	1	1

$$T_3 = Q_1 + \overline{Q_2}\,\overline{Q_3} + Q_2Q_3$$

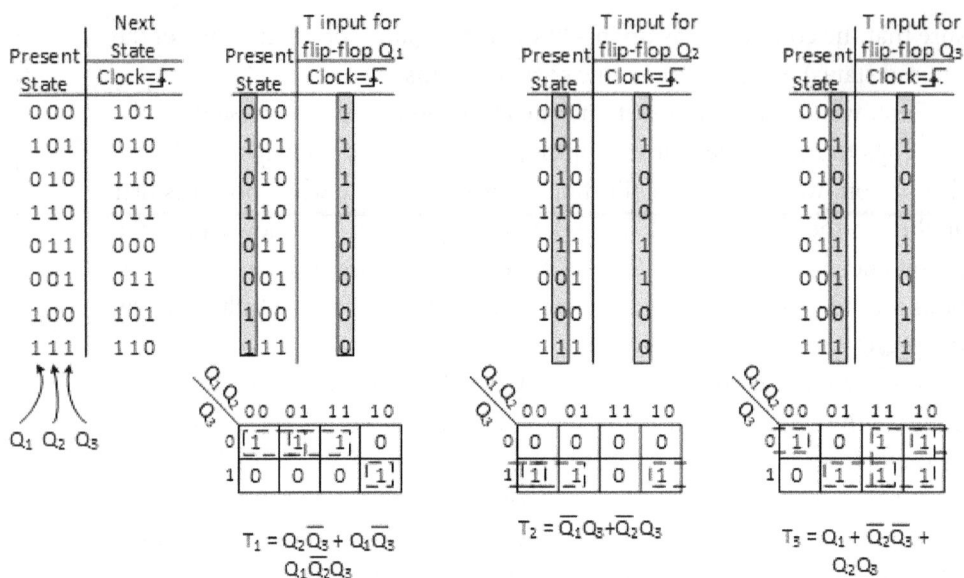

FIGURE 8.14. Logic diagram of a complex 5 state counter.

self-starting counter is the same as for designing any counter. The only additional step that is involved consists of using the unused states to transition to one of the used states. In our example, we have transitioned from the three unused states to three different states. This is not necessary; the transitions could be to any state or even to another unused state, as long as the machine eventually transitions to one of the used states.

Review Questions for Section 8.1.2

Question: The counter in Figure 8.8 uses six states. It has two unused states. Modify the counter so that it is a self-starting counter.

Answer: One possible self-starting state transition diagram, along with the state transition table, are shown in Figure 8.15. There are many others that are possible.

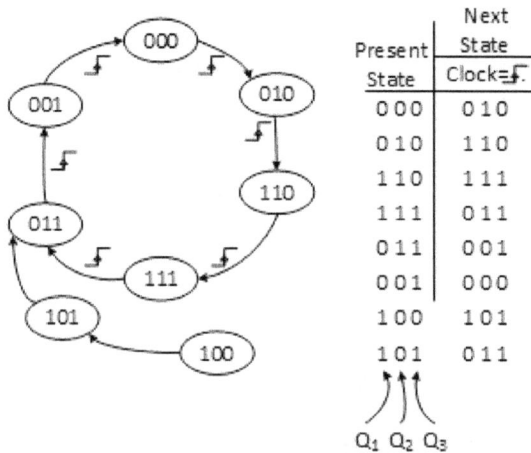

Present State	Next State Clock=⌐
0 0 0	0 1 0
0 1 0	1 1 0
1 1 0	1 1 1
1 1 1	0 1 1
0 1 1	0 0 1
0 0 1	0 0 0
1 0 0	1 0 1
1 0 1	0 1 1

$Q_1\ Q_2\ Q_3$

FIGURE 8.15. Modifying the Transition Table of a 3-bit counter to make it self-starting.

Question: Draw the excitation table for all three flip-flops, assuming that the D flip-flop is to be used for all three flip-flops in the self-starting counter.

    Answer: The excitation tables for all three flip-flops are shown in Figure 8.16.

Question: If a counter has to count fourteen distinct counts or events, what is the minimum number of flip-flops required to build this counter?

    Answer: When a counter uses one flip-flop, it can count up to two different counts. With two flip-flops, it can count up to four different states. With three flip-flops, it can count up to eight different states. With four flip-flops, it can count up to sixteen different states. So our counter would need four flip-flops. With four flip-flops, there would be fourteen states in the count sequence and two states which would be unused.

8.2. FINITE STATE MACHINES

These machines are very similar to the counters that we studied in the previous section. The similarity arises from the fact that the counters and the state machines can exist only in specific states. The difference between the two is

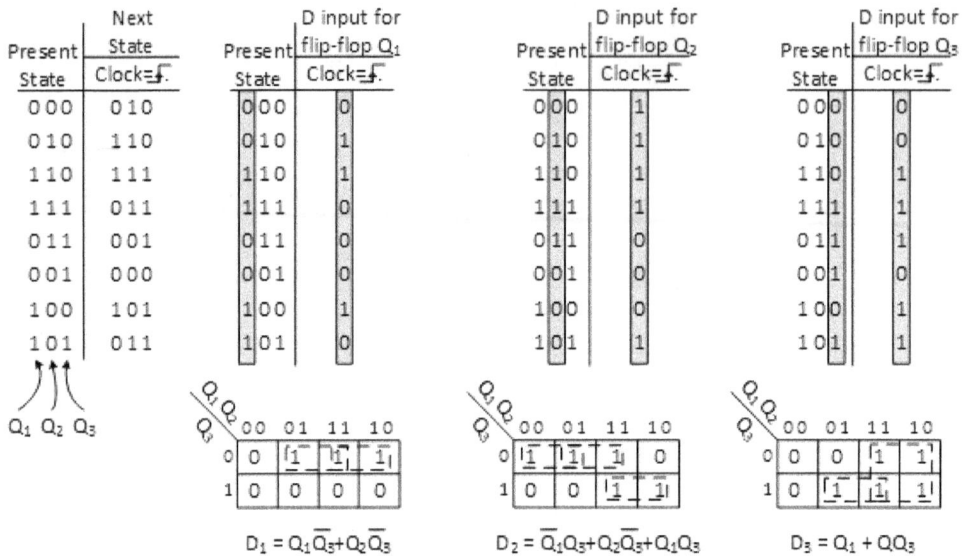

Present State	Next State Clock=↓		Present State	D input for flip-flop Q_1 Clock=↓		Present State	D input for flip-flop Q_2 Clock=↓		Present State	D input for flip-flop Q_3 Clock=↓
0 0 0	0 1 0		0 0 0	0		0 0 0	1		0 0 0	0
0 1 0	1 1 0		0 1 0	1		0 1 0	1		0 1 0	0
1 1 0	1 1 1		1 1 0	1		1 1 0	1		1 1 0	1
1 1 1	0 1 1		1 1 1	0		1 1 1	1		1 1 1	1
0 1 1	0 0 1		0 1 1	0		0 1 1	0		0 1 1	1
0 0 1	0 0 0		0 0 1	0		0 0 1	0		0 0 1	0
1 0 0	1 0 1		1 0 0	1		1 0 0	0		1 0 0	0
1 0 1	0 1 1		1 0 1	0		1 0 1	1		1 0 1	1

$Q_1 Q_2 Q_3$

$D_1 = Q_1\overline{Q_3} + Q_2\overline{Q_3}$

$D_2 = \overline{Q_1}Q_3 + Q_2\overline{Q_3} + Q_1Q_3$

$D_3 = Q_1 + QQ_3$

FIGURE 8.16. Converting the three bit counter to self starting counter.

minor. While the counter has only the clock pulse as the input, the finite state machines have one or more external inputs in addition to the clock input. The transition in the counter occurs on the clock pulse. The transition in the state machine occurs when a clock pulse arrives and the inputs are in a specific condition. The counters visit all the states in sequence. The states visited by the finite state machines are not in any specific order: the next state in a finite state machine depends on the present inputs. Generally, the outputs and the next state of a finite state machine are combinational logic functions of both the inputs and present state. Since the next state of these finite state machines depends on the value of the present input to the machine, these machines can be more complex than the counters. These machines are invaluable in building circuits that are used for control and decision-making logic in digital systems.

While counters are also finite state machines, there are some real differences between the two. First, a counter as a machine goes through a predefined sequence of states. This sequence never changes: with each clock pulse, the counter always advances to the next state. A finite state machine, on the other hand, does not go through a predefined sequence of states. The machine does transition from one state to the next with every clock pulse, but the next state that the finite state machine transitions to is determined by the present input to the machine. Another difference between the counter and a finite state machine is that the counter does not need an external input (other than a clock pulse), while a finite state machine depends on its input to determine the next state that it will go to. A third difference is in the method of describing the behavior. A counter is described by the sequence of states that it has to transition to, while a finite state machine is described by a word description, which the designer has to interpret and convert to a state transition diagram.

8.2.1. A SIMPLE FINITE STATE MACHINE

We will begin our study of finite state machines with a simple example. The purpose of this machine is to count the number of logic high inputs that arrive on a digital input that is a one-bit input line. We do not want to count the number of logic highs; instead, we wish to know if we have had an odd number of logic high inputs or an even number of logic high inputs. This is referred to as the *parity* of the binary input stream.

From the description of this machine, we can determine that the machine will have two states: one state that tells us that we have received an even number of logic high inputs, and a second state that tells us that we have received an odd number of logic high inputs. The machine will start out in the even state, so on reset we want the machine to begin in the even state. This machine will change states every time we have a logic high input. With this understanding, we can draw the state transition diagram for the machine, as shown in Figure 8.17.

To draw the state diagram, we first begin with the idea that when we have received no logic high inputs, the machine tells us that we have received an even number of logic high inputs. That is why we have labeled the reset state as "even." In this state, we have to inform the user, with an output from the machine, that the machine has received an even number of logic high inputs. The output is indicated under the label used for the state. In this case, it is a zero. To determine what happens to the machine, we assume that the machine will receive an input when the machine is in the "even" state. This input will be either a zero or a one. If the input is a zero, then the machine stays in the same state, as it has received an even number of logic inputs

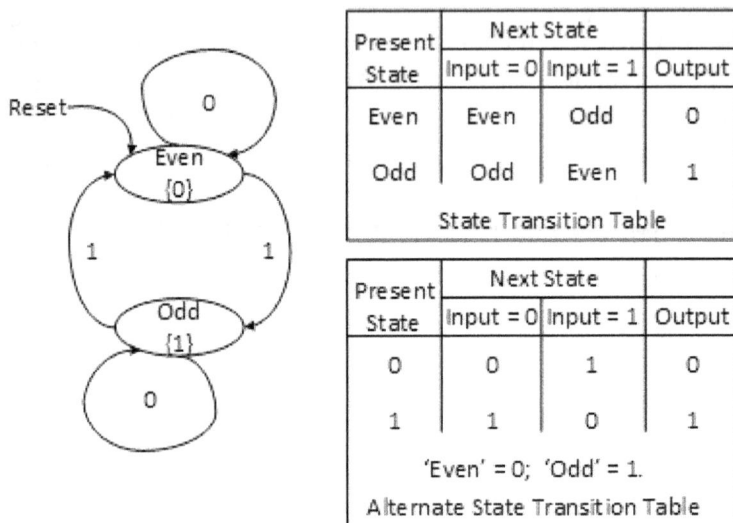

Present State	Next State		Output
	Input = 0	Input = 1	
Even	Even	Odd	0
Odd	Odd	Even	1

State Transition Table

Present State	Next State		Output
	Input = 0	Input = 1	
0	0	1	0
1	1	0	1

'Even' = 0; 'Odd' = 1.

Alternate State Transition Table

FIGURE 8.17. A 2 state finite state machine.

so far. This transition is indicated by a loop around to the same state, with a zero within the loop. This indicates that the machine moves from state "even" to state "even." This indicates the transition that the machine makes when the input to the machine is a zero. The second input to the machine is a one. If the input is a one, then the machine transitions to state "odd," as this new input would make the number of logic high inputs odd. This is indicated by an arrow going from the state "even" to the state "odd," with a one written next to it. The arrow indicates the transition that the machine makes, and the number written next to the arrow tells us what the input should be for the machine to make the specific transition.

The "odd" state is the second state of the machine. When the machine is in this state, it has received an odd number of logic high inputs. Since the machine has received an odd number of inputs, the output from this state is a one. This is written inside the state, under the name of the state. This informs the user that an odd number of logic high inputs have been received. When the machine is in the "odd" state, it will receive an input which will be either a zero or a one. If the input is a zero, then the machine stays in the same state. This is indicated by a loop around to the same state, with a zero within the loop. This indicates the transition that the machine makes when the input to the machine is a zero. The second input to the machine is a one. If the input is a one, then the machine transitions to state "even." This is indicated by an arrow going from the state "odd" to the state "even," with a one written next to it.

With this, we have transformed the word description into a state transition diagram. The next step is to build a state transition table. We transform the state transition diagram to the state transition table, just as we did when we were designing the counters. The state transition table for our machine is also shown in Figure 8.17. This time, note that we have several differences. First, we have an external input, and the state transitions depend on the external input. The transition table shows this by having a column for each possible input the machine can receive in every state. The transitions from the state "even" are to state "even" when the input is zero, and the transition from state "even" to state "odd" when the input is a one. The second difference is that now we have an output column. This column tells us what the output of the machine is when the machine is in any particular state. In the counter, the state of all the flip-flops was considered as the output of the machine.

To continue with the design of the machine, we have to assign binary values to all the states in the machine. We do this so that we can use flip-flops and logic components to build the state machine. This is also done in Figure 8.17. For this machine, we have made the state assignment of "even" = 0 and "odd" = 1. This is a simple replacement, so everywhere we have the state "even" in the state transition diagram, we replace it with a zero, and everywhere we have the state "odd" in the state transition diagram, we replace it with a one. In general, the idea is to assign a unique combinations of zeros and ones to each state. We have to make sure that no two states have the same state assignment. Other than that, there are no rules for a state assignment.

Now we can choose any type of flip-flop to build this machine. Let us decide that we will use the T flip-flop to build the machine. Once the state assignment is complete and we have chosen the type of flip-flop to be used, the next step is to determine the excitation table and the excitation equation for the T flip-flop. This is done in Figure 8.18. Notice that the procedure for getting the excitation equation is the same for a finite state machine as it was for the counters that we designed in the previous section. In addition to the excitation equation for the flip-flops, we also have an output logic for this machine. The equation for the output logic for the machine is obtained

State Transition Table

Present State	Next State In = 0	In = 1
0	0	1
1	1	0

Present State	Output
0	0
1	1

Out = Q_1

Excitation Table

Present State	Next State In = 0	In = 1
0	0	1
1	0	1

Q_1 / In

	0	1
0	0	0
1	1	1

$T_1 = In$

In → T_1 Q_1 → Out

Clr

Reset

Logic diagram of the state machine

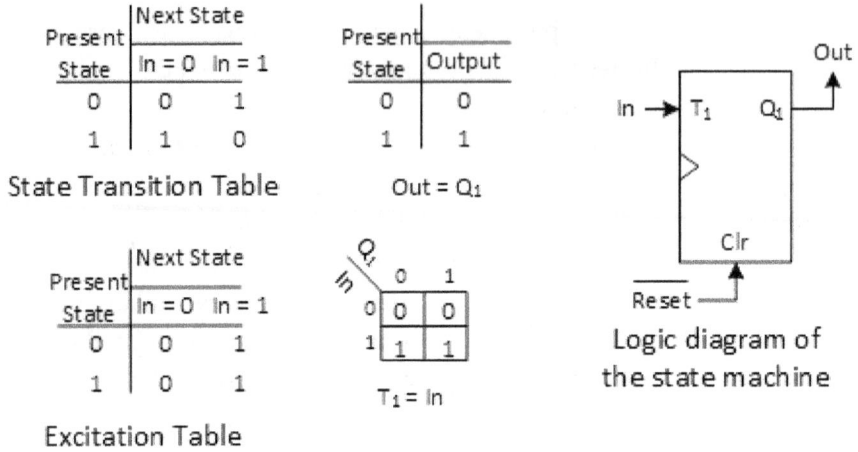

FIGURE 8.18. Determining the excitation and output for the state machine.

using its own K-map; we have also included the output equation for this machine in Figure 8.18. The implementation of the finite state machine is shown in Figure 8.18.

8.2.2. THE BASIC APPROACH

The design procedure that we followed in the previous section to design a finite state machine can be divided into four basic steps. We list them here and give some details about each step.

Step 1. *Interpret the problem from the word description:* The finite state machine is usually described by listing conditions that have to be satisfied. You should understand these conditions and determine the unique sets of conditions for each state and for each transition. Each state of the finite state machine will represent one such condition that has to be satisfied. The machine may also have some secondary conditions. Represent them in an unambiguous manner.

Step 2. *Represent the machine in a state transition diagram:* From the list of conditions given, build a state transition diagram.

When doing this, assign a name to each state, and try to use names that reflect the condition that the state is satisfying. Along with the name of the state, also specify the output that the machine is supposed to give out when the machine is in that state. From the state transition diagram, build a state transition table. To build the state transition table, you will need to know how many flip-flops are required to build the machine. The number of flip-flops can be determined by using Equation (8.1)

$$\text{Number of States} \leq 2^{\text{Number of Flip-flops}} \qquad (8.1)$$

Step 3. *Minimize the function and obtain the logic equation:* The first step is to minimize the machine. Sometimes when we build the state transition table, we may have redundant states. Since these are redundant states, we will identify and remove them. After the machine is minimized, each of the states is assigned a unique combination of the flip-flop outputs. This is known as *state assignment*. Once the states are assigned a unique combination of flip-flop outputs, we build the excitation table for each flip-flop that is to be used to build the machine.

Step 4. *Implement the finite state machine:* Using the excitation tables, we determine the excitation equation for each input to all the flip-flops, along with a unique equation for the output. Once we have the equations, we can build the logic that will give us the state machine that goes through the states as different inputs arrive with each clock pulse.

The next example will use the steps laid out above to build a finite state machine. We begin with a word description. Let us assume that we wish to build a machine that has one input and one output. The input to the machine is a sequence of single bits. The purpose of the machine is to recognize when a particular sequence of consecutive inputs has been received. We will also assume that the machine receives a new input every clock pulse. The output from the machine is always zero unless the machine has identified the input sequence (… 011 …) on three consecutive clock pulses. When this event occurs, the output changes to one for one clock pulse. The output will return to zero, and will stay at zero until this sequence is detected again. A typical sequence of input and output is shown in Equation (8.2)

$$\text{input} = 0\,0\,1\,0\,0\,1\,\underbrace{0\,1\,1}_{1}\,0\,0\,\underbrace{0\,1\,1}_{1}\,0\,1\,0\,1\,\underbrace{0\,1\,1}_{1}0001 \qquad (8.2)$$

Equation (8.2) shows a typical sequence of inputs. With each clock pulse, the output will be a zero, except during the last bit of the three sequences that have been identified. The output will be a zero for the first two bits, but during the time of the last bit, the output will be a one. With this understanding, we can say that the machine is to detect the input sequence 011. When the machine finds this sequence, it will output a one on the third bit. Otherwise, the machine will always output a zero. This completes Step 1 in the design procedure of our state machine.

To complete Step 2 of our design, we will follow the state transition diagram shown in Figure 8.19. In the state transition diagram, we begin with the *"reset"* state. In this state, we have received no part of the sequence, and the output is a zero. Both of these are shown in the circle that we are using to represent the state. When the machine is in this state, it may receive an input of "1." Since a "1" is not the start of the sequence that we have to detect, we say that the machine has received no input that the machine has to remember. In this case, the machine transitions back to the reset state. In the reset state, the machine may receive an input of a "0." This input is the start of the sequence that the machine has to detect. We have to tell the machine that it has to remember this input. To remember any unique event,

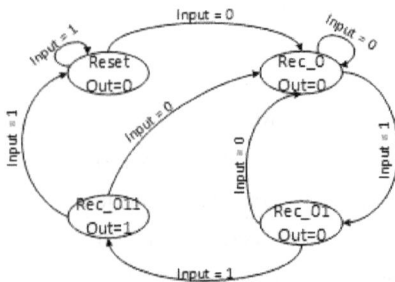

State transition diagram

Reset   = 0 0
Rec_0   = 0 1
Rec_01 = 1 0
Rec_011 = 1 1

State Assignment

State transition table

Present State	Next State		
	Input = 0	Input = 1	Output
0 0	0 1	0 0	0
0 1	0 1	1 0	0
1 0	0 1	1 1	0
1 1	0 1	0 0	1

FIGURE 8.19. State transition diagram and State transition table for sequence detector.

we take the machine to a new state. A state remembers what the last event was. This time we want to remember that we have received an input of "0." We have also labeled the state as "*Rec_0*" as an aid to remember what this state is supposed to represent. The output in this state is a zero. Since the "*reset*" state can only have inputs of a "0" or a "1," we have finished accounting for the "*reset*" state. Next, we go and see how the machine behaves in the new state that we have created.

When the machine is in "*Rec_0*" state, it will receive an input of either a "0" or a "1." If it receives an input of "0," then the last two inputs have been "0 0." Since only one zero is part of the sequence that we have to remember, we can forget the first zero and return the machine back to state "*Rec_0*," where it will remember that the last input was a "0." When the machine receives an input of "1," then the last two inputs have been "0 1." Since this is part of the sequence that the machine has to remember, we must take the machine to a new state so that it will remember the last inputs. The new state the machine will go to is labeled "Rec_*01*," where it will remember that the last two inputs were "0 1." The output in this state "*Rec_0*" is a zero since we have not yet completed the sequence. Since the "Rec_0" state can only have inputs of a "0" or a "1," we have finished accounting for the "*Rec_0*" state. We next go and see how the machine behaves in the new state that we have created.

When the machine is in "*Rec_01*"state, it will receive an input or either a "0" or a "1." If it receives an input of "0,"then the last three inputs have been "0 1 0." With this input, the sequence that we have to detect is broken. From this sequence, only the last zero is part of the sequence that we have to remember, and we can forget the previous inputs. To remember the last "0," we return the machine back to state "*Rec 0*," where it will remember the last "0." When the machine receives an input of "1," then the last three inputs have been "0 1 1." Since this is part of the sequence that we have to remember, we must take the machine to a new state so that it will remember this. The new state the machine will go to is "*Rec_011*," where it will remember the last three inputs of "0 1 1." The output in the state "*Rec_01*" is a zero, since we have not yet completed the sequence. Since the "*Rec_01*" state can only have inputs of a "0" or a "1," we have finished accounting for the "*Rec_01*" state. Next, we go and see how the machine behaves in the new state that we have created.

When the machine is in "*Rec_011*" state, it will receive an input or either a "0" or a "1." If it receives an input of "0," then the last four inputs have been "0 1 1 0." This represents a start of a new sequence. From this new sequence, only the last zero is part of the sequence that we have to remember; we can forget the previous inputs. To remember the last "0," we return the machine back to state "*Res_0*," where it will remember the last "0." When the machine receives an input of "1,"then the last four inputs have been "0 1 1 1." This does not represent the start of a sequence after the previous sequence is completed. Since this is not part of the sequence that we have to remember, we take the machine to the "*reset*" state so that it will remember that we have received no part of the sequence. The output in the state "*Rec_011*" is a one, since we have completed the sequence.

This completes the state transition diagram. In the state transition diagram (as an aid to memory), we have used names for the states. Now, as we are preparing to determine the excitation equations for the flip-flops, we must assign binary values to the states. Since we have four states, according to Equation (8.1), we can use two flip-flops. So we have made the assignment for the states. Each state can be assigned any binary combination that we like. The only condition is that no two states may have the same binary combination. One possible state assignment is shown in Figure 8.19. Once the binary assignment is made, we can draw a state transition table. This is also shown in Figure 8.19. This finishes Step 2 of the design procedure.

Step 3 requires a little explanation. This step consists of two different actions. The first step is to minimize the state transition table. Why do we need to minimize the state transition table? We do this to reduce the number of states in the state machine. When we have fewer states in a machine, the hardware logic required to build that machine will very likely have fewer gates. Therefore, it helps to minimize the state transition table. There are several different techniques to minimize the state transition table. Here we will look at only the simplest method, shown in Figure 8.20, to minimize the state transition table.

In any state transition table, when two states have identical outputs (and have transactions to identical states for all input conditions), then the two states can be merged into one. Merging the two states into one involves removing any reference to one of the states in the entire state transition table. In place of the removed state we substitute the other state which is its equivalent.This way, we have only one of the two states remaining. For an

| Present | Next State | | Output |
State	Input = 0	Input = 1	
0 0 0	1 1 0	0 0 0	0
0 1 0	0 1 0	0 1 1	0
1 0 0	0 1 0	0 1 1	0
1 1 0	0 1 1	1 0 0	1
0 1 1	0 1 1	1 0 0	0

Merge these two
states

Cannot Merge
these two states

FIGURE 8.20. Mergine states together in a State transition table
by merging states together.

example, look at Figure 8.20. In the state transition table, we see that states
(0 1 0) and (1 0 0) have the same output, and both of the states transition
to state (0 1 0) when the input is zero, and to state (0 1 1) when the input is
one. This meets all the criteria for merging the two states into one state. To
merge the two states, we replace all occurrences of state (0 1 0) in the state
transition table with the state (1 0 0). On the other hand, states (1 1 0) and
(0 1 1) both transition to state (0 1 1) when the input is zero, and they both
transition to state (1 0 0) when the input is a one. These two states meet
one condition to be merged together into one state. Their output, however,
is not the same. State (1 1 0) has a one output, while state (0 1 1) has a zero
output. Since the output from the machine is different in the two states, the
two states cannot be merged into one state. Merging the states together into
one state is that simple. First, check the outputs; if they are the same, then
check the transitions. If both the outputs and the transitions are the same,
then the states can be merged into one state. For this particular example,
when we merge the two states (0 1 0) and (1 0 0) into one state, we will
have only four states left over in our state machine. Without merging the
states together, we would need three flip-flops; after we merge the states into
one, we have only four states left over. With only four states, we can reduce
the number of flip-flops to two. Minimizing the state transition table can
be as simple as that. Another method of minimizing the state transition
table is given in Section 8.4 in this chapter. Okay, back to the example we
were working on.

Examine the state transition table in Figure 8.19. We see that states (0 0)
and (1 1) have identical transactions when the input is zero, and also when
the input is a one. These two states, however, have different outputs, so we

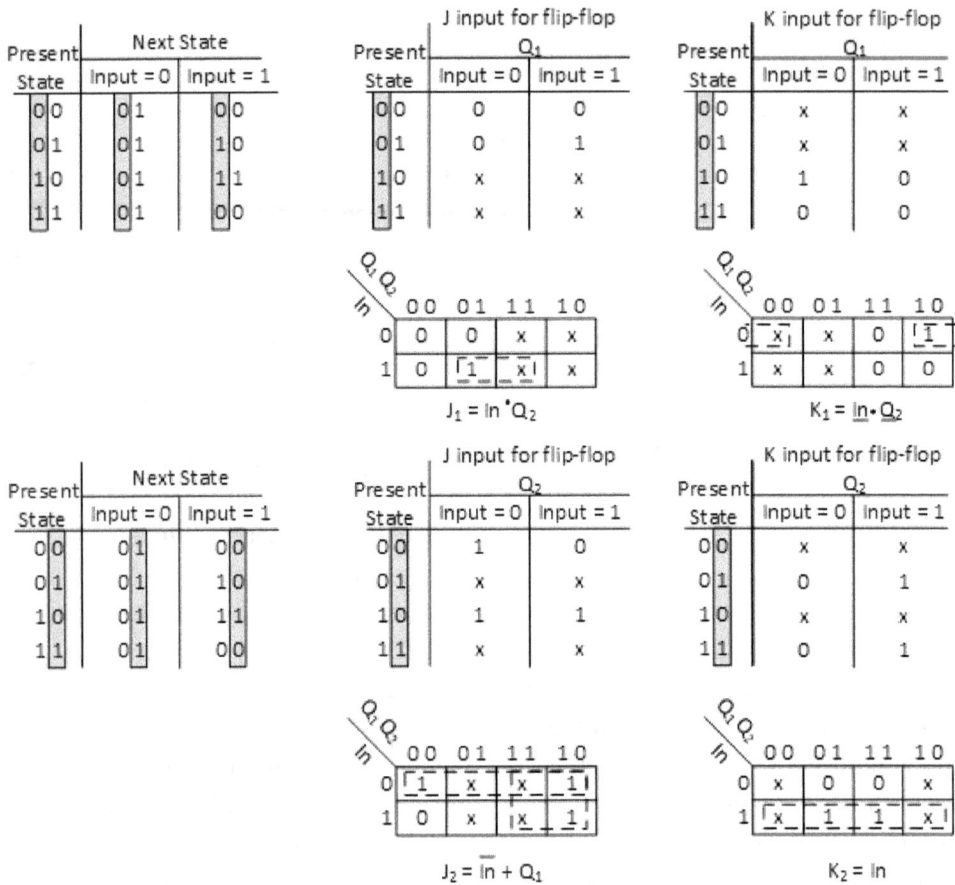

Present State	Next State	
	Input = 0	Input = 1
00	01	00
01	01	10
10	01	11
11	01	00

J input for flip-flop Q_1

Present State	Input = 0	Input = 1
00	0	0
01	0	1
10	x	x
11	x	x

K input for flip-flop Q_1

Present State	Input = 0	Input = 1
00	x	x
01	x	x
10	1	0
11	0	0

$Q_1 Q_2$ / In

In \ Q_1Q_2	00	01	11	10
0	0	0	x	x
1	0	1	x	x

$$J_1 = In \cdot Q_2$$

In \ Q_1Q_2	00	01	11	10
0	x	x	0	1
1	x	x	0	0

$$K_1 = \overline{In} \cdot Q_2$$

Present State	Next State	
	Input = 0	Input = 1
00	01	00
01	01	10
10	01	11
11	01	00

J input for flip-flop Q_2

Present State	Input = 0	Input = 1
00	1	0
01	x	x
10	1	1
11	x	x

K input for flip-flop Q_2

Present State	Input = 0	Input = 1
00	x	x
01	0	1
10	x	x
11	0	1

In \ Q_1Q_2	00	01	11	10
0	1	x	x	1
1	0	x	x	1

$$J_2 = \overline{In} + Q_1$$

In \ Q_1Q_2	00	01	11	10
0	x	0	0	x
1	x	1	1	x

$$K_2 = In$$

FIGURE 8.21. Determining the excitation for the sequence identifier.

cannot merge these two states into one state. We can say that the state transition table we have represents a minimum machine. For this machine, we can determine (using Equation 8.1) that we will have to use two flip-flops. Say we use the J K flip-flops: our next step is to determine the excitation for the flip-flops. This is done in Figure 8.21.

The last step in designing the finite state machine is to build the logic diagram that will give us the state transition machine from the logic equations that we have obtained for the machine in Figure 8.21. This is done in Figure 8.22. This completes the design of the sequence identifier.

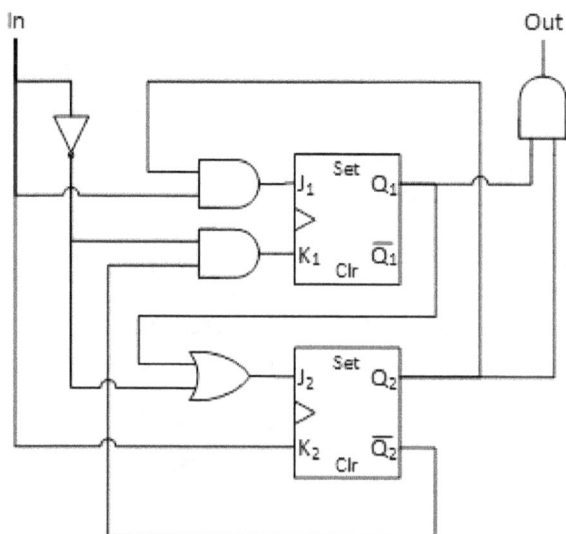

FIGURE 8.22. Logic diagram of the sequence identifying state machine.

8.3. MOORE AND MEALY SEQUENTIAL MACHINES

The sequential machine that we designed in Section 8.2 had its output determined by the state of the machine. When the machine got to state "even" in Figure 8.17, it had an output of "0." It did not matter how the machine got there. The same was true for the other state "odd" in that machine. When the machine got to a state, its output was known. This type of a design of a sequential machine is known as a *Moore* design of a sequential machinemachine. In a Moore design, the output is associated with the state of the machine. An alternate way to design a sequential machine would be to assign the output to the transition that the machine takes. This design of a sequential machine, where the output of the machine is associated with the transition, is known as the *Mealy* design. In a Mealy design, the output is associated with the transition that the machine makes when it receives an input. A state transition diagram and a state transition table for a typical Mealy design of a sequential machine are shown in Figure 8.23.

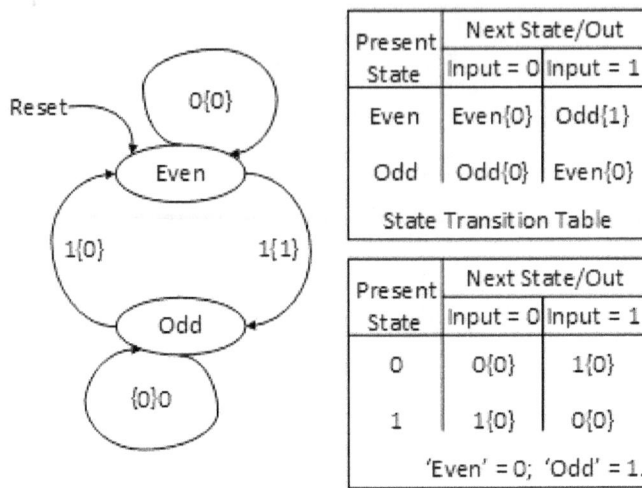

Present State	Next State/Out	
	Input = 0	Input = 1
Even	Even{0}	Odd{1}
Odd	Odd{0}	Even{0}

State Transition Table

Present State	Next State/Out	
	Input = 0	Input = 1
0	0{0}	1{0}
1	1{0}	0{0}

'Even' = 0; 'Odd' = 1.

FIGURE 8.23. Mealy state machine.

A simple Mealy machine (shown in Figure 8.23) has one input and one output. Each transition edge is labeled with the value of the input and the value of the output (the output is inside the curly brackets). The machine in Figure 8.23 implements the same state machine as the one we examined in Figure 8.17. Notice that there is not much difference in how we get the state transition diagram and the state transition table for either the Moore or the Mealy machine. In a Mealy machine, since the output is associated with the transition, the output is listed with the transition, and not in the state. Our machine starts out in the state "even." When it receives an input of (0), it transitions to state "even," and the output during this transition is a (0). When the machine is in state "even" and receives an input of (1), it transitions to state "odd," and during this transition, the output is a (1). The output stays at the level that it is at until the next transition. Even though the transition is almost instantaneous, the output stays at the level until the start of the next transition.

Let us look at the state transition diagram and the state transition table of a second simple Mealy machine, shown in Figure 8.24. This machine can be thought of as an *Exclusive Or* machine: its output is the "Exclusive Or" of the last two inputs. If the last two inputs are the same, then the output is a zero, and if the last two inputs are different from each other, then the output is a one. When the machine is in the "reset" state, it has not received any

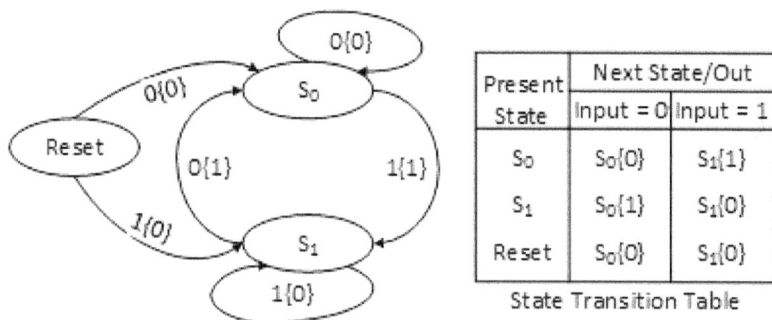

FIGURE 8.24. Exclusive-Or Mealy state machine.

input. In this state, it will receive an input of either a zero or a one. Upon receiving a zero, the machine goes to state S_0, and upon receiving a one, the machine goes to state S_1. Since this is the first input to the machine, both the transitions give a zero output.

In state S_0, the last input to the machine was a zero. If, in state S_0, the machine receives a zero, then the machine transitions from state S_0 to state S_0; during this transition, the output is a zero, as the last two inputs were both zero. If, in state S_0, the machine receives a one, then the machine transitions from state S_0 to state S_1; during this transition, the output is a one, as the last two inputs were a zero followed by a one.

In state S_1, the last input to the machine was a one. If, in state S_1, the machine receives a one, then the machine transitions from state S_1 to state S_1; during this transition, the output is a zero, as the last two inputs were both one. If, in state S_1, the machine receives a zero, then the machine transitions from state S_1 to state S_0; during this transition, the output is a one, as the last two inputs were a one followed by a zero. This is shown in both the state transition table and the state transition diagram in Figure 8.24.

Let's examine a third example of a Mealy machine. This time, we will reexamine the example of the sequence detector that we studied in Figure 8.19. The sequence detector was designed to detect the sequence (0 1 1) and give an output of (1) when the sequence was detected. The state transition diagram and the state transition table are given in Figure 8.25. The excitation tables for the flip-flops for the sequence detector are given in Figure 8.26. The entire design process, after drawing the state transition diagram, is the same for both the Mealy and Moore state machines. As a general rule, the Mealy machine will have fewer or the same number of

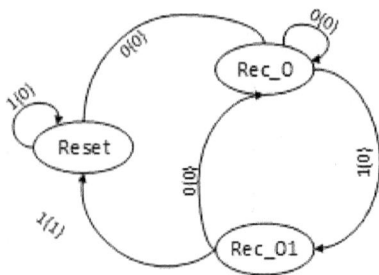

State Assignment

Reset = 0 0
Rec_0 = 0 1
Rec_01 = 1 0

Present State	Next State/Out Input = 0	Input = 1
0 0	0 1{0}	0 0{0}
0 1	0 1{0}	1 0{0}
1 0	0 1{0}	0 0{1}

State transition diagram

State transition table

FIGURE 8.25. Sequence detector as a Mealy Machine.

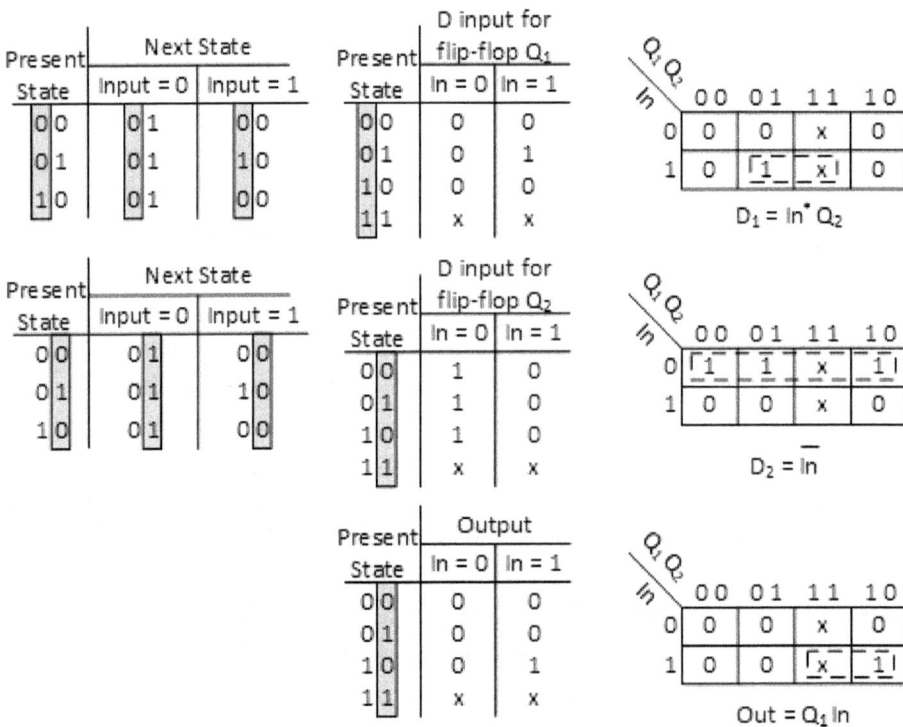

Present State	Next State Input = 0	Input = 1
0 0	0 1	0 0
0 1	0 1	1 0
1 0	0 1	0 0

Present State	D input for flip-flop Q_1 In = 0	In = 1
0 0	0	0
0 1	0	1
1 0	0	0
1 1	x	x

$D_1 = In \cdot Q_2$

Present State	Next State Input = 0	Input = 1
0 0	0 1	0 0
0 1	0 1	1 0
1 0	0 1	0 0

Present State	D input for flip-flop Q_2 In = 0	In = 1
0 0	1	0
0 1	1	0
1 0	1	0
1 1	x	x

$D_2 = \overline{In}$

Present State	Output In = 0	In = 1
0 0	0	0
0 1	0	0
1 0	0	1
1 1	x	x

$Out = Q_1 \, In$

FIGURE 8.26. Determining the excitation for the sequence identifier.

states as a Moore machine. The Mealy machine will never have more states than a Moore machine. In a Moore machine, there is no direct connection from the input to the output. Since there is no direct connection between the input and the output, the output does not have the input as part of the output equation. In a Mealy machine, there is a direct connection from the input to the output. Since there is a direct connection between the input and the output, the output will have the input variable as part of the output expression. This can be seen in Figure 8.26 for a Mealy machine, and in Figure 8.21 and Figure 8.22 for the sequence detector.

Review Question for Section 8.3

Question: Draw the state transition diagram and the state transition table for a machine that will have an output of one when the sequence 010 or the sequence 110 is detected. This machine has to be the Mealy design.

Answer: The state transition diagram and the state transition table are given in figure 8.27.

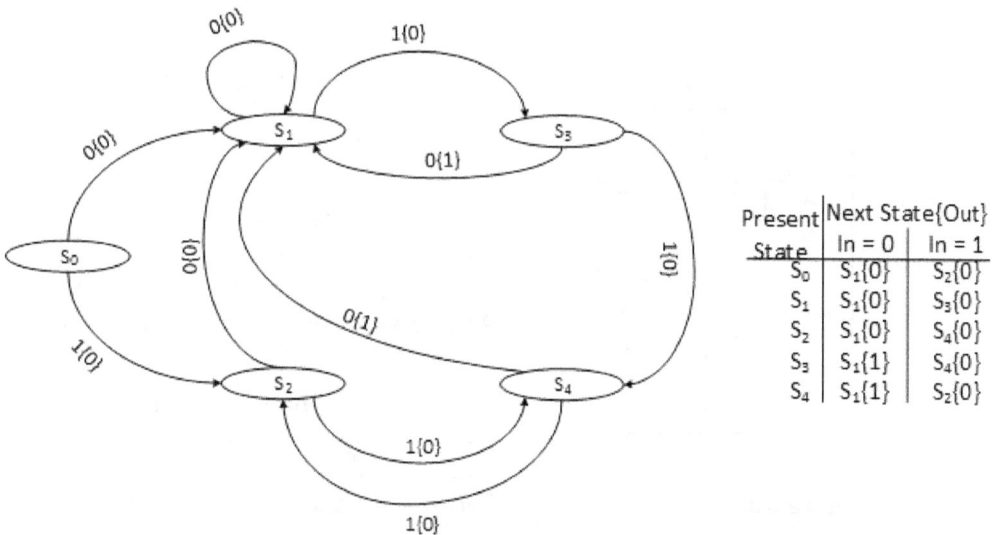

Present State	Next State{Out} In = 0	In = 1
S_0	$S_1\{0\}$	$S_2\{0\}$
S_1	$S_1\{0\}$	$S_3\{0\}$
S_2	$S_1\{0\}$	$S_4\{0\}$
S_3	$S_1\{1\}$	$S_4\{0\}$
S_4	$S_1\{1\}$	$S_2\{0\}$

FIGURE 8.27. Sequence detector to detect either (0 1 0) or (1 1 0).

8.4. TECHNIQUES FOR MINIMIZING A SEQUENTIAL MACHINE

Here we will discuss the *partitioning method* to minimize the state transition table. The main idea in minimizing a state transition table is to identify two or more states that will give us the same result (final output and state) from the machine if the machine starts in any one of these states. In other words, if the machine is in a black box and we provide it with a sequence of inputs, then we will not be able to tell, just by looking at the output, which one of the group of states the machine started in. If this is the case, then all the states in this group of states can be merged into one state. To explain this method, we will use the state transition table shown in Figure 8.28. This method works by portioning states such that all the states in each partition can be merged into one state.

The first partition is always according to the output from the states. In our state transition table, we have states {A, B, C, E, and F} that have an output of "0;" call this group of states "partition α." The other states {D, G, and H} have an output of "1;" call this group of states "partition β." This is our first partition, and it is shown in Figure 8.28. The states from two

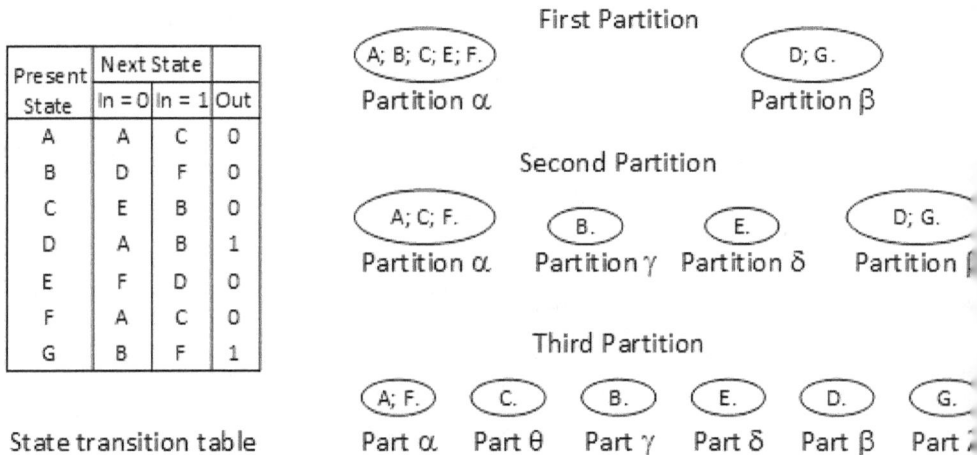

Present State	Next State		Out
	In = 0	In = 1	
A	A	C	0
B	D	F	0
C	E	B	0
D	A	B	1
E	F	D	0
F	A	C	0
G	B	F	1

State transition table

First Partition

(A; B; C; E; F.)   Partition α   (D; G.)   Partition β

Second Partition

(A; C; F.) Partition α   (B.) Partition γ   (E.) Partition δ   (D; G.) Partition [

Third Partition

(A; F.) Part α   (C.) Part θ   (B.) Part γ   (E.) Part δ   (D.) Part β   (G.) Part

FIGURE 8.28. State transition table to show the partitioning method of minimizing a sta machine.

different partitions can never be merged into one state. None of the states in partition α will ever be merged with any of the states in partition β.

When the machine is a Moore design, the output is associated with the state, and the partition is easily obtained. When the machine is a Mealy design, the output is associated with the transition that the machine makes. To make the partitions in a Mealy machine, we must match the output from each of the transitions from states.

Next, we check the transitions from each of the states. If two states transition to the same partition for the same input, then they belong together in one partition. For our example, when we check the transitions of state A, we find that it transitions to partition α when the input is "0," and to partition β when the input is a "1." This is also true for state C and state E. State B, however, transitions to partition α when the input is "0," and again to partition α when the input is a "1." Thus, the transitions of state B are not the same as the transitions of the states {A, C, and E}. Since the transitions are not to the same partition, we separate state B from its group of states and place it in a new partition. When we check the transitions of state F, we find that state F transitions to partition α when the input is "0," and again to partition α when the input is a "1." This is the same as state B, so we put both of these states in their own partition. Call this partition "Partition γ."

We continue this process of checking the transitions for the states in the other partition. In partition β, all the states transition to partition α when the input is "0," and again to partition α when the input is a "1." Since all the states in this partition transition to the same partitions, we leave this partition as it is. The result of checking the transitions is given in Figure 8.28 as the second partition. Now that we have new partitions, we can throw away the previous partitions and work with only the new partitions.

Since we broke one partition into two smaller partitions, we have to do the checking all over again for all the states in all the partitions. We begin by checking the transitions in partition α. All the states {A, C, and E} in this partition transition to partition α when the input is "0," and to partition β when the input is a "1." This partition stays as it is for now. When we check the transitions in partition γ, we find that all the states {B and F} in this partition transition to partition α when the input is "0," and to partition γ when the input is a "1." This partition also stays as it is for now. When we check the transitions in partition β, we find that the states {D and G} in this partition transition to partition α when the input is "0," and to partition γ when the input is a "1." State {H}, on the other hand, transitions to partition γ when the input is "0," and to partition α when the input is a "1." These are not the same transitions as the other two states in this partition. Since the transitions are not the same, we have to separate state {H} into its own partition. We have called this partition δ. The result of all this checking is shown in Figure 8.28 as the third partition. With these new partitions, we can forget the partitions that we had in the second partition.

Since we broke one partition into two smaller partitions, we have to do the checking all over again. This time we are looking at four partitions. On checking the transitions, we find that no more divisions of any of the partitions are necessary, since all the states in each of the partitions transition to the same partition. This means that all the states in partition α can be merged into one state. The same applies for all the states in partition β, and for partition γ and for partition δ. Since we have four partitions, we will end up with a state transition table with only four states.

Review Question for Section 8.4

Question: A state transition table is given in Figure 8.29. Use the partitioning method to determine the minimum state machine.

    Answer: Figure 8.29 also shows the steps followed in partitioning the state machine. This time, we can see even before we begin that states A and F have identical transitions, so they can be merged together into one state. Going through the partition method, we find that these are the only two states that can be merged together into one state.

8.5. CHAPTER PROBLEMS

8.5.1.   Draw the state transition diagram of a counter that counts according to the sequence given: $00001 \rightarrow 00010 \rightarrow 00100 \rightarrow 01000 \rightarrow 10000 \rightarrow 01001 \rightarrow 00101 \rightarrow 00011 \rightarrow 00001$.

8.5.2.   The counter in problem 8.5.1 is a non-self-starting counter. Modify the diagram to make it into a self-starting counter.

Present State	Next State In = 0	In = 1	Out
A	A	D	0
B	C	F	0
C	E	G	0
D	A	B	1
E	C	D	0
F	A	B	0
G	E	F	1
H	B	A	1

State transition table

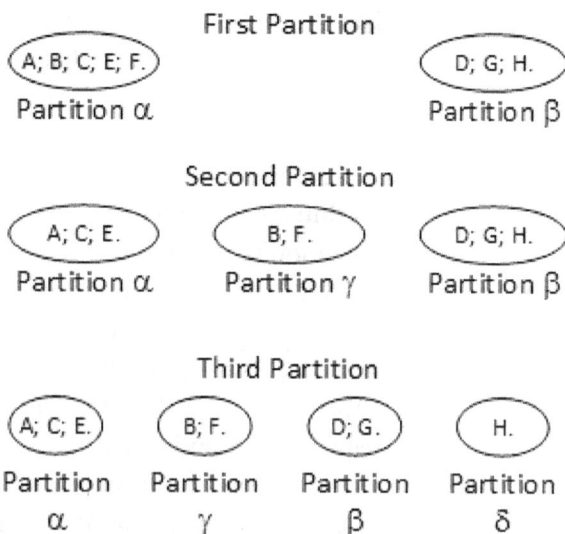

First Partition

A; B; C; E; F.   Partition α

D; G; H.   Partition β

Second Partition

A; C; E.   Partition α

B; F.   Partition γ

D; G; H.   Partition β

Third Partition

A; C; E.   Partition α

B; F.   Partition γ

D; G.   Partition β

H.   Partition δ

FIGURE 8.29. State transition table to show the partitioning method of minimizing a state machine.

8.5.3. Draw the state transition diagram and the state transition table of a counter that counts according to the sequence given: $001 \rightarrow 010 \rightarrow 100 \rightarrow 110 \rightarrow 111 \rightarrow 011 \rightarrow 101 \rightarrow 001$.

8.5.4. The counter in problem 8.5.3 is a non-self-starting counter. Modify the diagram and the table to make it into a self-starting counter.

8.5.5. Determine the excitation equation for the counter in problem 8.5.3 if the flip-flop Q_1 is a D flip-flop, flip-flop Q_2 is a T flip-flop, and flip-flop Q_3 is an S-R flip-flop.

8.5.6. Draw the logic diagram for the counter in Problem 8.5.3.

8.5.7. Draw the state transition diagram and the state transition table of a counter that counts all the even numbers of a four-bit binary number.

8.5.8. Determine the excitation equation for the counter in problem 8.5.7 if the flip-flop Q_1 is a D flip-flop, flip-flop Q_2 is a T flip-flop, flip-flop Q_3 is a JK flip-flop, and flip-flop Q_4 is an S R flip-flop.

8.5.9. Draw the state transition diagram and the state transition table of a counter that counts all the odd numbers of a four-bit binary number.

8.5.10. Determine the excitation equation for the counter in problem 8.5.9 if the flip-flop Q_1 is a D flip-flop, flip-flop Q_2 is a T flip-flop, flip-flop Q_3 is a JK flip-flop, and flip-flop Q_4 is an S R flip-flop.

8.5.11. Draw the state transition diagram and the state transition table of a finite state machine that will detect the sequence 0101. Allow repeat in the sequence so that an input of … 00101011 … will show two occurrences of the sequence 0101.

8.5.12. Determine the number of flip-flops that are needed to build the machine. Assuming that all the flip-flops are D type, determine the excitation table and the logic diagram of this machine.

8.5.13. Draw the state transition diagram and the state transition table of a finite state machine that will detect the sequence 010 or the sequence 101. Allow repeat in the sequence so that an input of … 00101011 … will show four matches of the sequence (two for the sequence 010, and two for the sequence 101).

8.5.14. Determine the number of flip-flops that are needed to build the machine. Assuming that all the flip-flops are T type, determine the excitation table and the logic diagram of this machine.

8.5.15. Using the partition method, determine if the machine that you have built in problem 8.5.14 is a minimum machine. If it is not, then minimize the machine and build the minimum machine.

8.5.16. A finite state machine has two inputs and one output. Input x_1 represents a trigger input. So, when the input x_1 has received the input 11, the machine will start to monitor the input x_2. Now, if the input x_2 has the input sequence 110, there will be an output of one from the machine. Once the machine starts to monitor the input x_2, it will continue to monitor the input x_2 until the input x_1 receives the input 00. At this time, the machine will reset and wait for an input of 11 on the input x_1. Draw the state transition diagram and the state transition table. Build the machine using only J K flip-flops.

8.5.17. A finite state machine is to be designed so that it will follow the timing diagram shown in Figure P8.5.17. The diagram shows that the output has to be one when the input sequence received is $01 \rightarrow 10 \rightarrow 11 \rightarrow 11$.

FIGURE P8.5.17.